MINI Building

15 NEIGHBORHOOD FACILITY

CONTENTS

PARKLIFE	파크라이프	004
THE VEIL HOUSE	베일 하우스	018
NOBORI RESIDENTIAL BUILDING	노보리 빌딩	028
MUSEUM OF MODERN ALUMINUM THAILAND	태국 모던 알루미늄 박물관	040
FRENCH KITSCH Ⅲ CAFE	프렌치 키치 3 카페	052
PUBLIC LIBRARY IN BAIÃO, PORTO	포르투 바이앙 공공 도서관	062
TERRACE HOUSE	테라스 하우스	072
CITY FRESH	시티 프레시	086
FOUR-ROOF PAVILION	네 개의 지붕 파빌리온	098
MO288	엠오288	108
HOUSE AT TERUBOK	테루복의 주택	116
FLOP ART SPACE	플롭 아트 스페이스	124
CARRETAS	카레타스	136
WANSANTRAK COMMUNITY CENTER	완산뜨락 주민소통방	144
THE FLYING BLOCK HOTEL	플라잉 블록 호텔	154

QUAN SPHERE ㅣ 콴 스피어	**162**
ALIVE RESIDENCE ㅣ 얼라이브 레지던스	**174**
RESIDENCE KONGKAHERB ㅣ 레지던스 콩카허브	**182**
TILE NEST ㅣ 타일 네스트	**194**
HOUSE BELAKU ㅣ 하우스 벨라쿠	**204**
SECRET GARDEN ㅣ 비밀의 정원	**212**
HOUSE WITH AN EYE ㅣ 눈이 있는 집	**220**
ISAGI RESTAURANT ㅣ 이사기 레스토랑	**228**
THE AXOLOTL ㅣ 액솔로틀	**236**
HOUSE COVE(R) ㅣ 하우스 커버	**244**
CO-HOUSING DENVER ㅣ 코-하우징 덴버	**252**
APARTMENT S ㅣ 아파트 에스	**260**
THE HIÊN HOUSE ㅣ 히엔 하우스	**270**
PROFILE ㅣ 프로필	**280**

파크라이프
PARKLIFE

ARCHITECT : AUSTIN MAYNARD ARCHITECTS / ANDREW MAYNARD, MARK AUSTIN, MARK STRANAN

LOCATED IN BRUNSWICK AND SET WITHIN THE NIGHTINGALE VILLAGE - Australia's first carbon neutral residential precinct - ParkLife is a high-performing, beautifully designed, community-focused apartment building containing 37 homes and two commercial tenancies.

ParkLife has a distinctive mountainous roofline and a unique rooftop amphitheatre, as well as a variety of social/communal areas, diverse in scale, location and character. Homes within comprise of 14 one-bedroom, 19 two-bedroom, 2 three-bedroom and 2 Teilhaus apartments, each designed to extol space-efficiency, functionality and flexibility. Five of the apartments are designated social housing, through Housing Choices Australia.

The exterior of ParkLife is highly insulated white steel cladding, with cables, grills and rods to allow vegetation to proliferate. To reduce carbon and increase insulation performance we avoided using external pre-cast concrete panel, typically used in apartment buildings. The cladding's strong vertical lines softens the building to a residential scale, in opposition to standard cladding used in most apartment buildings.

The interior palette of materials were kept deliberately simple; timber floors, white walls, white cabinetry, concrete ceiling and a terrazzo tile in the bathroom. The intent was to allow the residents to personalise their homes rather than apply too many finishes and textures. Attractive, functional spaces were created with great views and natural light.

ParkLife deliberately encapsulates and celebrates design elements from some of our favourite residential projects over the years. Visual aspects such as the mountainous roof line, white steel balustrade and the colour yellow. The colour extends a visual consistency throughout the building, including all common area thoroughfares, bike store and large planter boxes, as wells as the articulate balconies on the Northern side. As a way of avoiding unnecessary expense (and in turn increasing the cost of each home), services within the common areas remain exposed. To avoid a messy environment full of different pipes and conduits we decided to paint everything in the same yellow, so the eye sees the space not the smaller details. The glazed staircase is also bright yellow and, rather than a standard dimly-lit concrete tunnel, is light-filled with sweeping views and greenery, while showcasing the life within the building.

Location Melbourne, Australia **Use** Housing **Site area** 706m² **Total floor area** 4,210m² **Completion** Novenber 2021 **Project manager** Fontic **Developer** Austin Maynard Architects **Builder** Hacer Group **Engineers** Structural & Civil - Irwin Consult (now WSP), Mechanical Engineer - Irwin Consult (now WSP), Electrical Engineer - Irwin Consult (now WSP), Hydraulic Engineer - Irwin Consult (now WSP), Fire Services Engineer - Irwin Consult (now WSP), Fire Safety Engineer - Irwin Consult (now WSP) **ESD** Irwin Consult (now WSP) **Traffic Consultant** GTA Consultants **Planning consultant** Hansen **Access Consultant** Access Studio **Landscape Architects** Openwork **Building Surveyor** Steve Watson & Partners **Cost Consultant** WT **Photographer** Tom Ross

← Exterior view   → Distinctive mountainous roofline

↑ Distinctive mountainous roofline & a unique rooftop amphitheatre

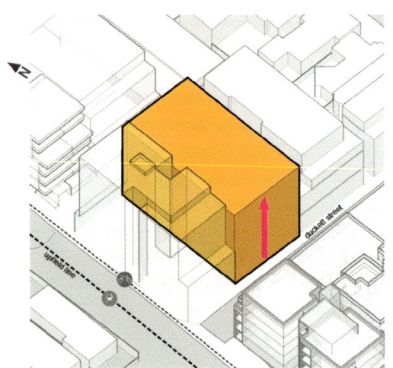

1. Seven Storey Massing

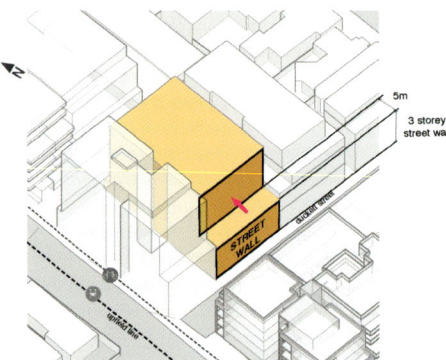

2. Street Wall

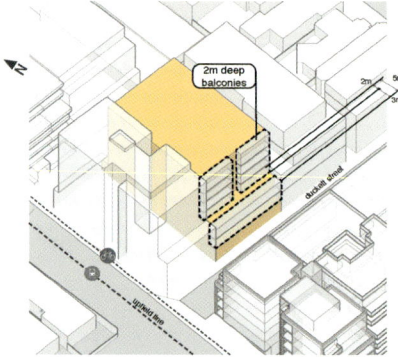

3. Street Wall and Balconies

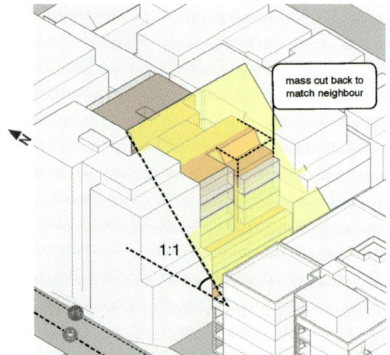

4. Mass Reduction

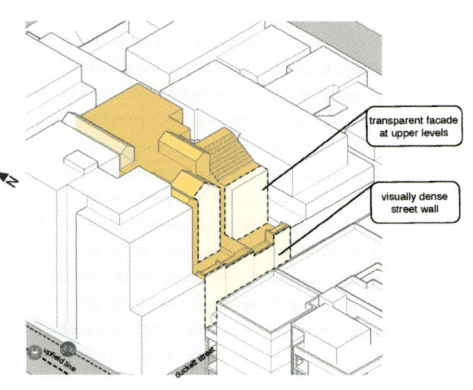

5. Street Wall Articulation

MASSING DIAGRAM

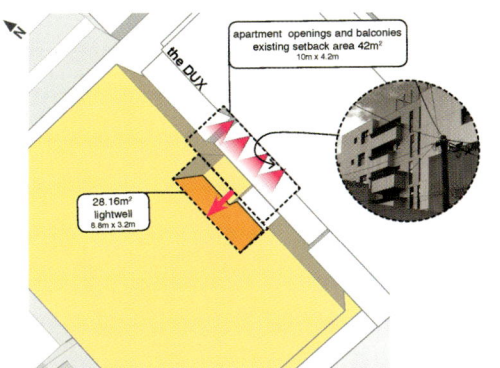

6. Interface with the Dux
Light-well setback to the balconies and living area windows of the dux

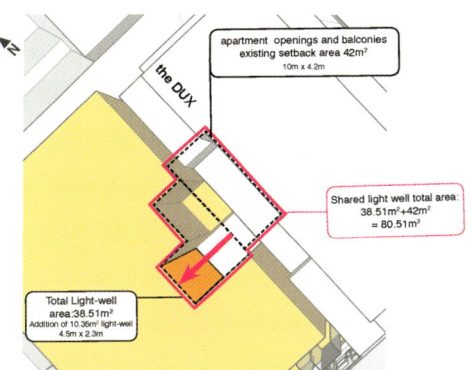

7. Interface with the Dux
Further light-well setback

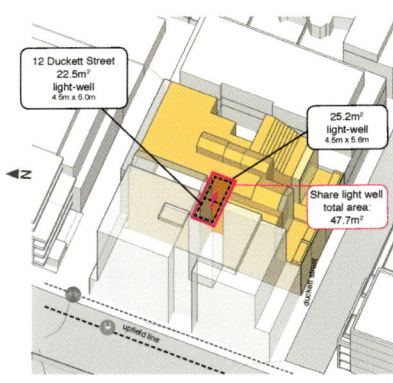

8. Shared Light-well with 12 Duckett Street
Generous Light-well setback to apartments, lift lobby and neighbours at both-sides

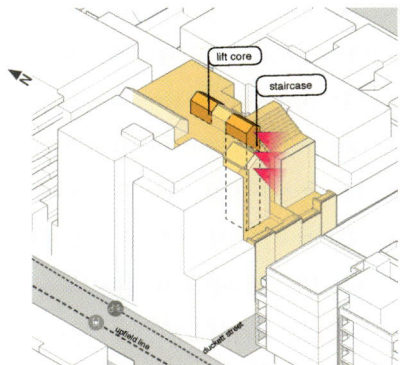

8. Views form the Staircase

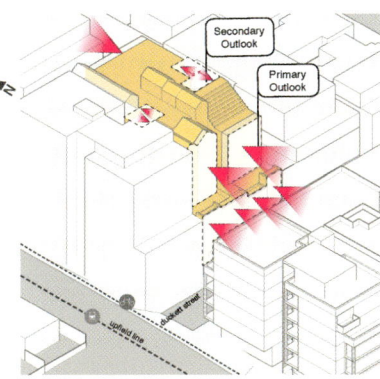

8. Primary and secondary Outlook

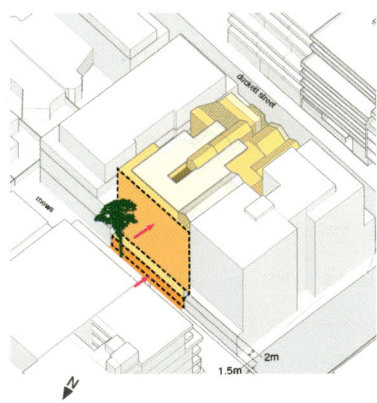

9. Mews and the Protecting Existing Tree
Based on advice from our arborist, refer to further diagrams & included report.

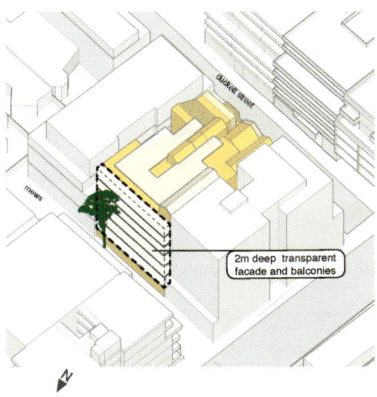

10. Rear Laneway the Dux
Balcony to construct around tree branches.

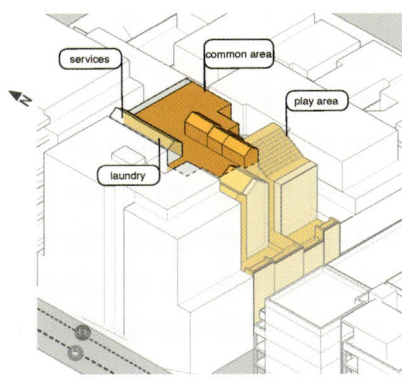

11. Communal Rooftop Space and Amenity
Refer to roof plan for further information.

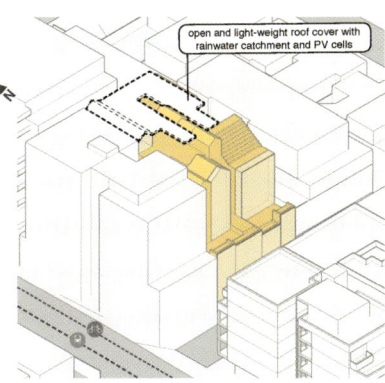

12. Communal Rooftop Space and Amenity

MASSING DIAGRAM

South facade view & white steel cladding

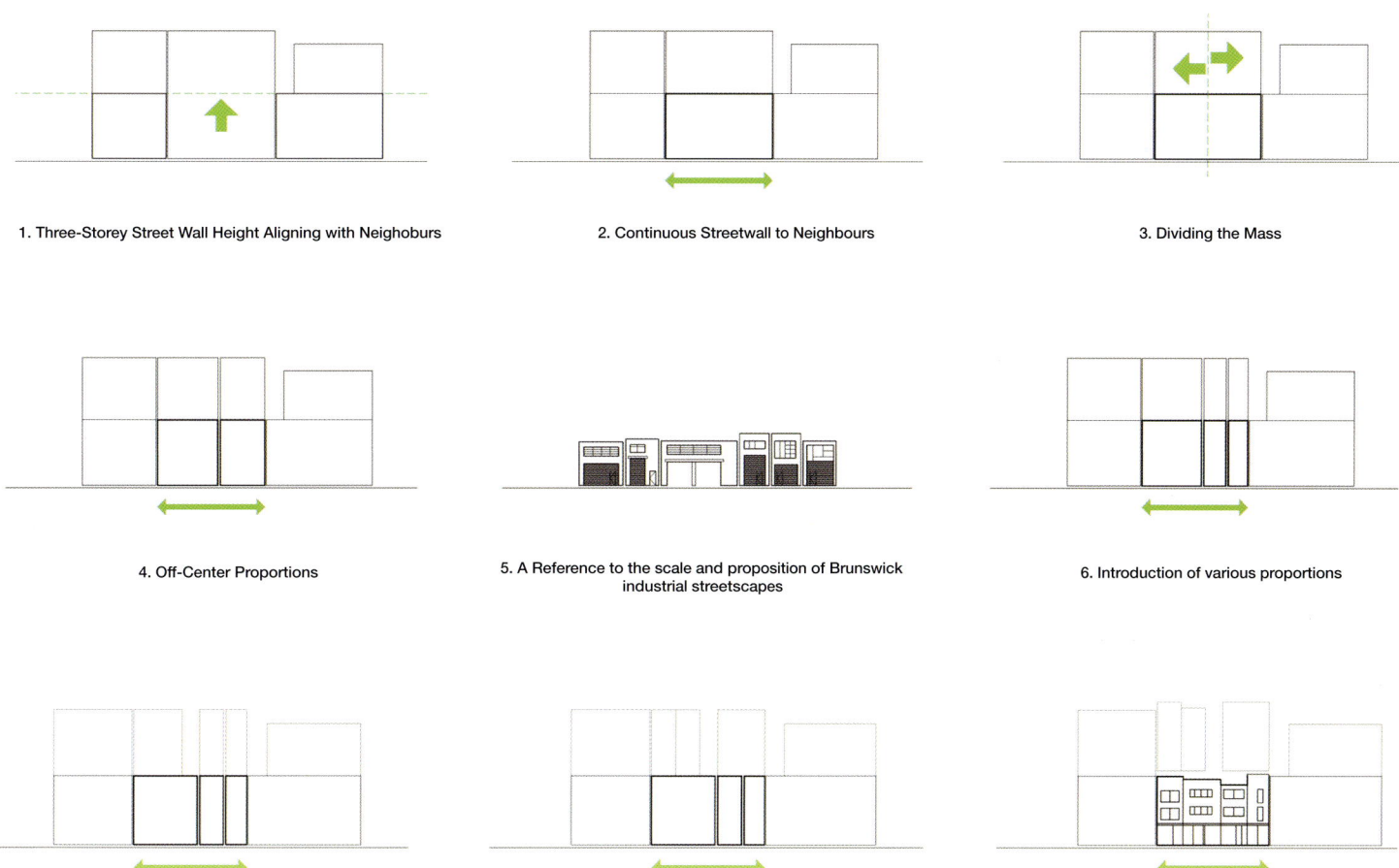

1. Three-Storey Street Wall Height Aligning with Neighoburs
2. Continuous Streetwall to Neighbours
3. Dividing the Mass
4. Off-Center Proportions
5. A Reference to the scale and proposition of Brunswick industrial streetscapes
6. Introduction of various proportions
7. Balance and Arrangement
8. Balance and Arrangement
9. Proportional Openings and Scale Reduction

SOUTH FACADE ARTICULATION DIAGRAM

브런즈윅에 위치한 파크라이프는 오스트레일리아 최초의 탄소 중립 주거 지구인 나이팅게일 빌리지 내에 있으며, 고성능과 아름다운 디자인의 커뮤니티 중심 아파트 건물로서 37개의 주택과 두 개의 상업 임대 공간을 포함하고 있다.

파크라이프는 독특한 산악 형태의 지붕선과 특별한 옥상 원형 극장을 자랑하며, 규모, 위치, 특성이 다양한 사회적/공동의 장소들을 제공한다. 이곳의 주택들은 14개의 원룸, 19개의 투룸, 2개의 쓰리룸, 그리고 2개의 털하우스 아파트로 구성되어 있으며, 공간 효율성, 기능성 및 융통성을 극대화하였다. 이 중 5개의 아파트는 하우스 초이스 호주를 통해 사회주택으로 지정되었다. 파크라이프의 외관은 단열 효과가 뛰어난 백색 강철 클래딩으로, 케이블, 그릴 및 막대가 식물을 번식하게 하였다. 탄소 배출을 줄이고 단열 성능을 향상하기 위해, 대부분의 아파트 건물에서 흔히 사용되는 외부 프리캐스트 콘크리트 패널 사용을 피했다. 클래딩의 강렬한 수직 라인은 대규모 아파트 건물에서 일반적으로 사용되는 표준 클래딩과 대조적 으로 건물을 주거용 규모로 부드럽게 만들었다.

주거 내부의 재료 팔레트는 의도적으로 단순하게 유지했다. 목재 바닥, 흰색 벽, 흰색 수납장, 콘크리트 천장 및 욕실의 테라조 타일 이러한 선택은 주거자들이 집을 개성 있게 꾸밀 수 있도록 하였으며, 너무 많은 마감재와 질감을 적용하는 대신에 주거자의 취향에 맞게 공간을 활용할 수 있다. 매력적이고 기능적인 공간은 멋진 전망과 자연 채광을 제공하였다.

파크라이프는 의도적으로 지난 몇 년간 우리가 선호하는 주거 프로젝트들로부터 영감을 받은 디자인 요소들을 포괄하고 기념한다. 산의 형태를 띤 지붕 라인, 흰색 강철 발코니 난간, 그리고 노란색 같은 시각적 요소들이 건물 전반에 걸쳐 일관된 시각적 연속성을 제공한다. 이는 모든 공용 구역 통로, 자전거 보관소 및 대형 식물 상자는 물론 북쪽에 위치한 발코니의 세련된 디자인에 이르기까지 확장된다. 불필요한 비용을 피하고 (그 결과 각 가정의 비용 증가를 방지하기 위해) 공용 구역 내의 설비는 노출되어 있다. 다양한 파이프와 전선관으로 지저분한 환경을 피하고자 모든 것을 동일한 노란색으로 도색하였다. 그래서 눈은 작은 세부 사항이 아닌 공간을 볼 수 있다. 또한, 유리로 된 계단도 밝은 노란색으로 칠해져 있으며, 일반적인 어두운 콘크리트 터널 대신 넓은 전망과 녹지로 가득 차 있으면서 건물 내부의 생명력을 보여준다.

↖ White steel balustrade & the colour yellow   ↑ Yellow balcony & planter box   ↗ Balcony large planter box

Upper Storey

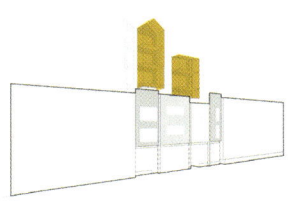

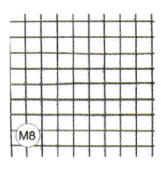
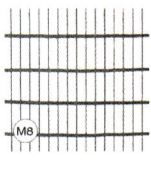
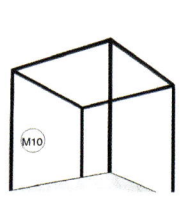

Galvanized Chain Link Mesh OR
Galvanized Metal Mesh
The use of full height chain-link or metal mesh with 100-120 mm aperatures provides a high degree of transparency.

Street Wall

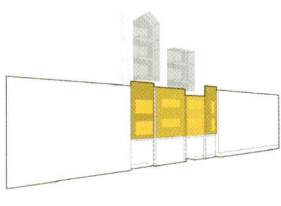

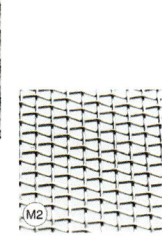
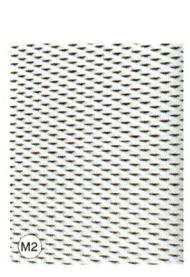
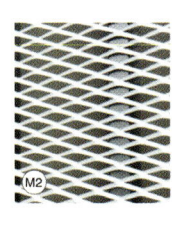

Web-forge OR
Expanded Metal Mesh
Visually dense material to provide a solid look and feel to assist in planning code compliance. The mesh allows light to penetrate through.

Cladding and Ground

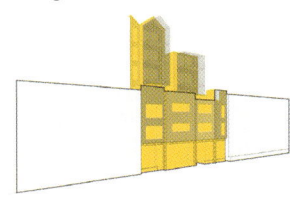

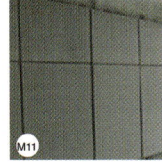

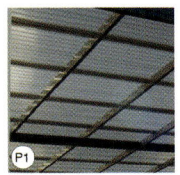
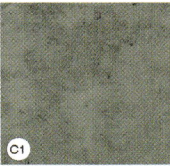

Metal Cladding, recycled brick, exposed concrete, powder coated yellow metal elements, and glazing.

FACADE MATERIALS

NORTH ELEVATION

SOUTH ELEVATION

SOUTH LIGHT COURT ELEVATION

NORTH LIGHT COURT ELEVATION

↑ External common area

↰ Yellow colour common area thoroughfares   ↱ Yellow colour common area thoroughfares   ← Yellow colour common area   → Bike store

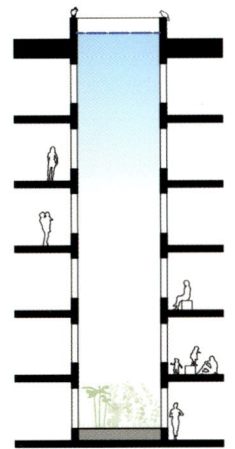

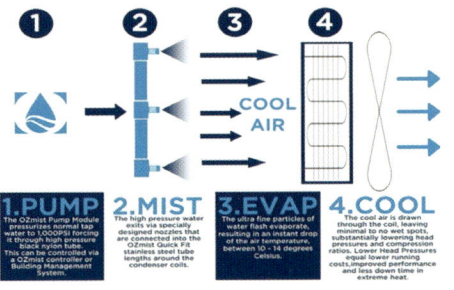

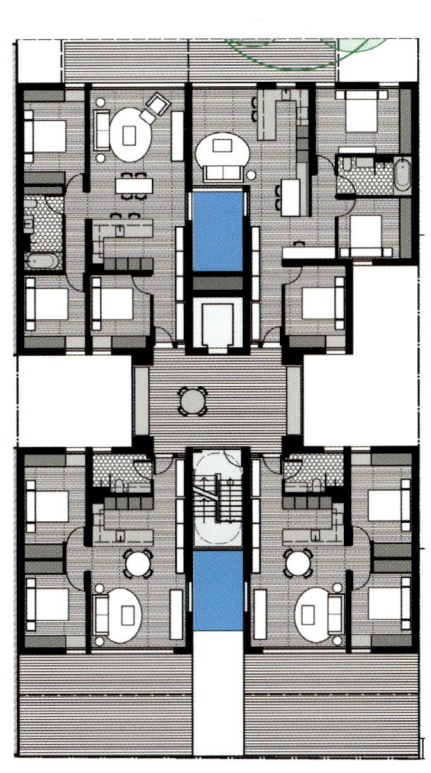

COOLING WORKS DIAGRAM

↑ Common area   ↵ Glazed staircase   ↳ Functional spaces & natural light.

← Great views and natural light → Great views and natural light ↙ Stair ↳ Living & dining room

← Unique rooftop　→ Distinctive mountainous roofline and a unique rooftop

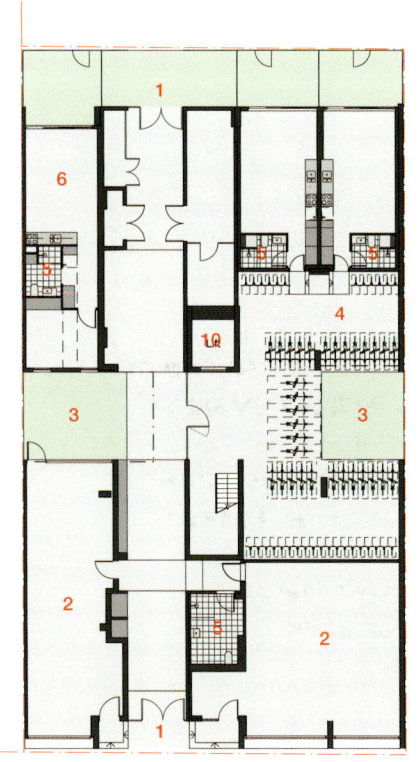

GROUND FLOOR PLAN

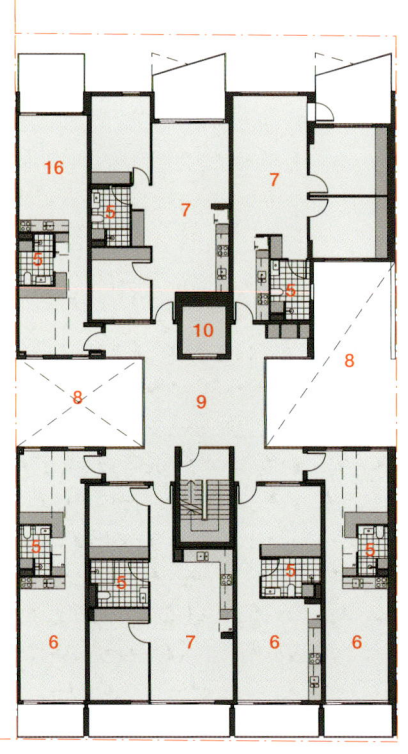

1ST FLOOR PLAN

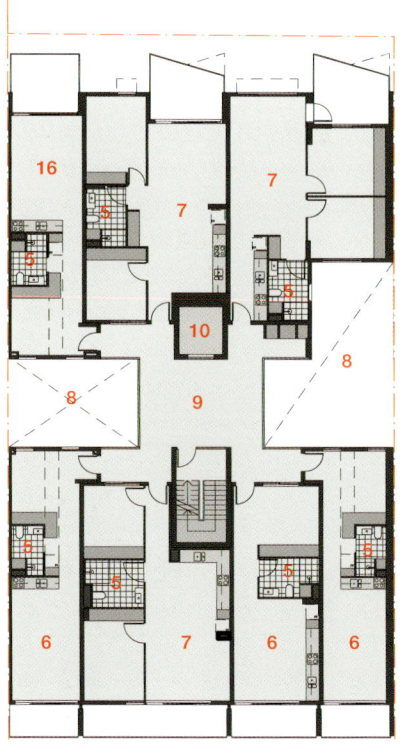

2ND FLOOR PLAN

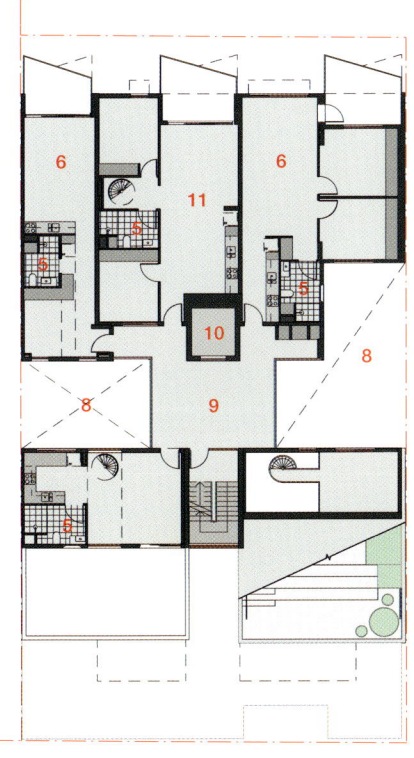
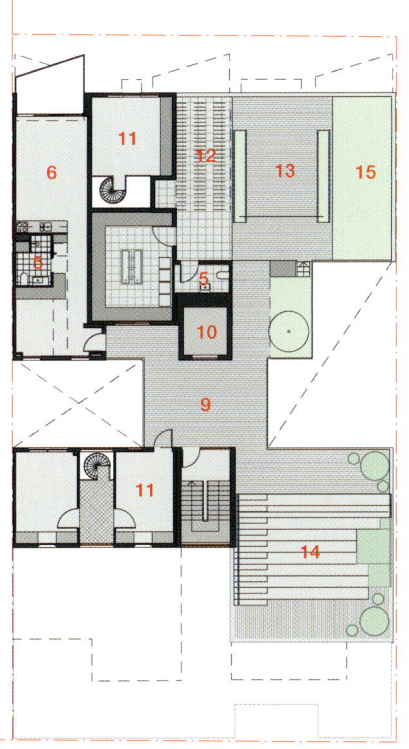

↑ Glazed staircase

6TH FLOOR PLAN **7TH FLOOR PLAN**

1	ENTRANCE	5	TOILET	9	COMMON AREA	13	COMMUNAL AREA
2	RETAIL	6	ONE BEDROOM	10	ELEVATOR	14	AMPHITHEATRE
3	GARDEN	7	TWO BEDROOM	11	TWO BEDROOM DUPLEX	15	LAWN
4	BIKE PARK	8	LIGHT COURT	12	HANGING AREA		

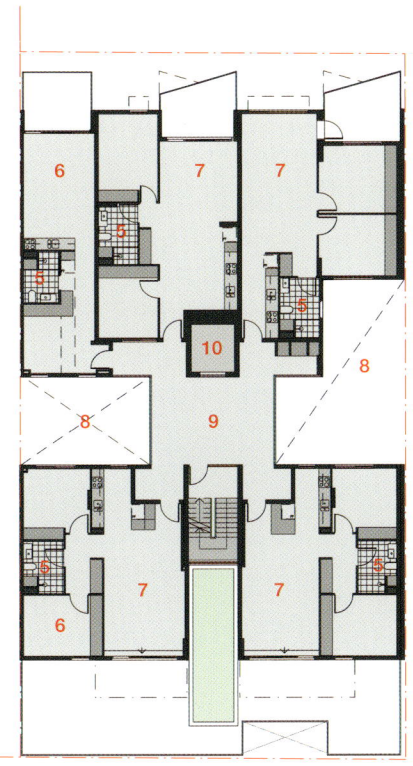
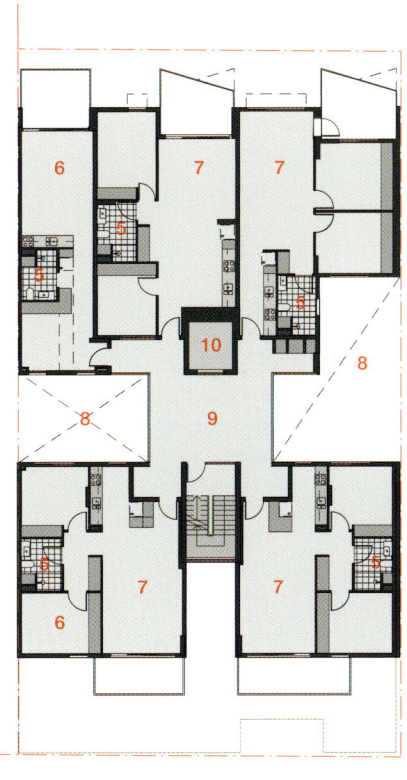
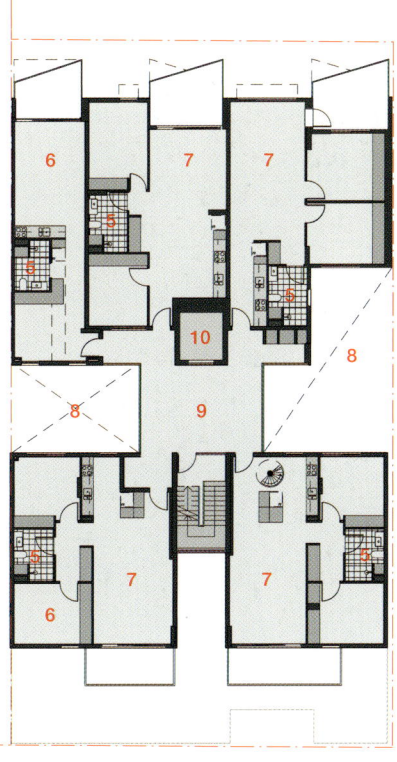

3RD FLOOR PLAN **4TH FLOOR PLAN** **5TH FLOOR PLAN**

베일 하우스
THE VEIL HOUSE

ARCHITECT : PAPERFARM INC / JARRETT BOOR, DANIEL YAO

SITUATED NEAR THE HISTORIC "TAIWAN-RENGA" (台灣煉瓦) brick kiln from 1899 that prospered this working-class district in Kaohsiung, Taiwan, the Veil House revisits this history by weaving a modern, tapestry-like facade utilizing floating clay bricks.

In a district with very narrow streets, close proximity to neighbors, and a hyperactive social fabric, privacy is often compromised. To maintain boundaries, windows are often shaded throughout the day; outdoor spaces, such as balconies and terraces, are left largely unused. The Veil House challenges this public/private dynamic of compact urban living by creating a peaceful retreat that redefines the nature of this neighborhood's typical house: a perforated brick facade liberates the need for window treatments and still allows filtered light into all the living spaces and bedrooms. The impetus for security and privacy reimagines the home as a body with a breathable, permeable skin. Like skin's pores, perforation density is devised according to the functional needs behind the enclosures.

The entry, through an interior garden, helps quiet the transition from the bustling city streets and provides a deep threshold into the heart of the home, thus acting as a type of perforation. The residents circle around an open atrium clad with 2 by 6 vertical aluminum louvers, to enter the main living area on the second floor. This materiality pays homage to another Taiwanese vernacular of protected fenestrations while enhancing the verticality of the home. Programmatically, this atrium is the engine of the house: it is an urban garden on the ground floor; on the bedroom's balconies it is a light-well introducing natural illuminance into the rooms; it is an airshaft for cross-ventilation with the brick veil at the front facade; and it is a connector that ties circulation and program together across multiple floors.

Throughout the home's interior, custom-designed terrazzo flooring defines spaces within the largely, open-plan living floors, while full-length, custom white-oak millwork conceals not only the kitchen but the entertainment and storage spaces as well. The reductive use of materials enhances the focus on the brick veil and the respite gained in the quiet, minimal interior. The desire to build a cozy, airy lifestyle behind an urban facade that successfully withdraws from the frenetic street life is the defining characteristic of the Veil House.

Location Kaohsiung, Taiwan **Use** Residence **Site area** 97m² **Building area** 65m² **Gross floor area** 315m² **Completion** 2023 **Project manager** Daniel Yao **Design team** Daniel Yao, Jarrett Boor, Bing-Yu Yu **Structural Engineer** Antop Structural Engineering **Contractor Team** Yuan-Shen Construction, Yih Tsae Interiors **Photographer** Daniel Yao

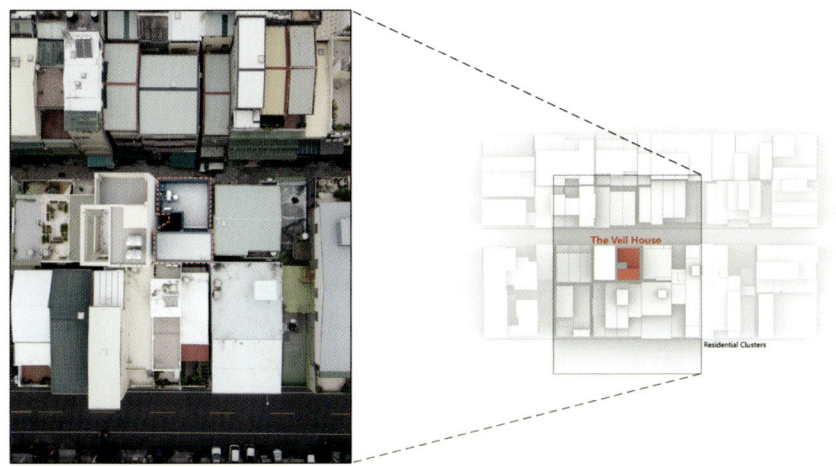

SITE PLAN

← Front facade

⌐ Side view   ⌐ Front view   ↑ Entrance

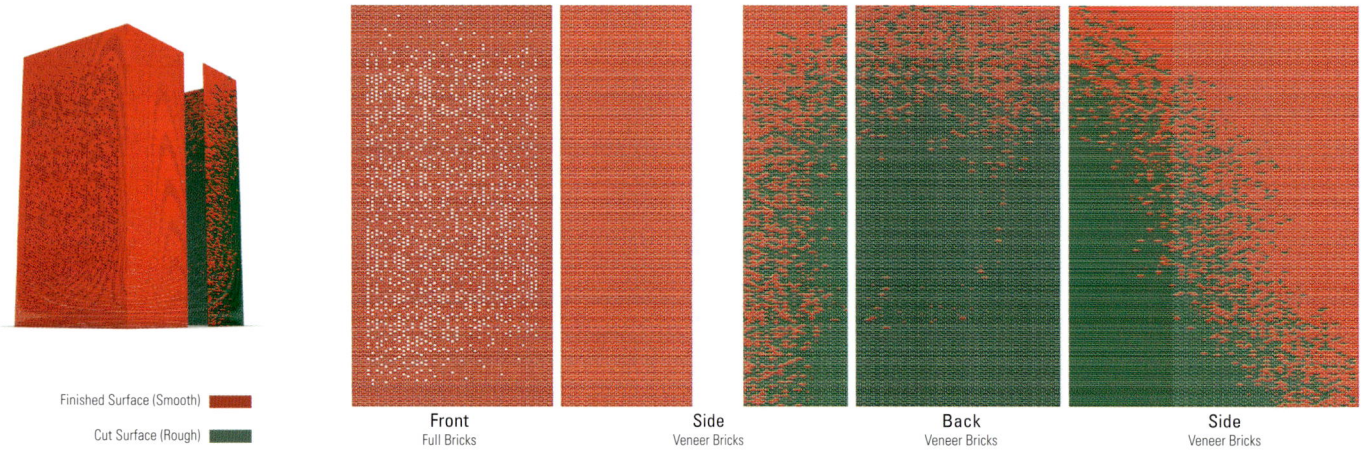

Finished Surface (Smooth) ■
Cut Surface (Rough) ■

Front
Full Bricks

Side
Veneer Bricks

Back
Veneer Bricks

Side
Veneer Bricks

BRICK SURFACE LAYOUT

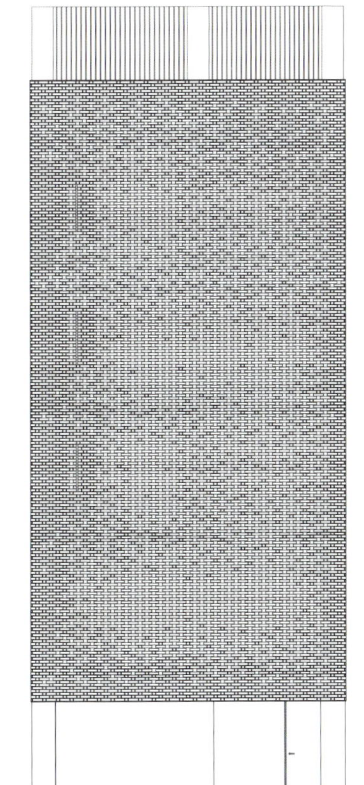

FRONT ELEVATION

SIDE ELEVATION

→ Night view

1899년 대만 가오슝(高雄)의 노동자 계급 지구를 번영시킨 유서 깊은 '대만-렝가'(台灣煉瓦) 벽돌 가마 근처에 위치한 베일 하우스는 부유하는 점토 벽돌로 현대적이고 태피스트리 같은 파사드를 짜내며 역사를 재조명한다.
매우 좁은 폭의 거리 및 이웃과의 근접성, 활발한 사회적 구조를 가진 지역에서는 종종 사생활이 침해된다. 창문은 경계를 유지하기 위해 창문은 하루 종일 가려져 있고, 발코니와 테라스와 같은 외부 공간은 대부분 사용지 않은 채로 자리한다. 베일 하우스는 도시 생활의 공공 및 사적 역학에 도전하며 이 지역의 전형적인 집의 본성을 재정의하는 평화로운 휴식처를 만든다. 천공된 벽돌 파사드는 창문 조성의 필요성에서 해방시키고 모든 거실과 침실로 여과된 빛을 충분히 허용한다. 보안과 사생활에 대한 원동력은 집을 통기성이 있고 투과성 있는 피부를 가진 몸으로 새롭게 상상하게 한다. 피부의 모공처럼 천공 밀도는 외피 뒤에 있는 기능적 필요에 따라 고안되었다.
내부 정원을 통한 입구는 번잡한 도시 거리로부터의 전환을 조용히 도와준다. 집의 중심으로 이어주는 깊은 문턱 역할을 하며, 일종의 타공으로 작용합니다. 거주자들은 2x6 수직 알루미늄 루버로 덮인 개방형 아트리움 주변을 돌아 2층의 주 생활 공간으로 들어갑니다. 이 물성은 집의 수직성을 향상시키면서 보호된 또 다른 대만에 경의를 표한다. 프로그램적으로 아트리움은 집의 엔진이다. 1층에는 도시 정원이 조성되어 있으며, 침실의 발코니에는 자연적인 조도를 방으로 도입하는 라이트 웰(light-well), 정면 파사드에는 벽돌 베일이 있는 교차 통풍을 위한 에어 샤프트(airshaft), 그리고 여러 층에 걸쳐 순환 및 프로그램을 연결하는 커넥터(connector)가 자리한다.
집 내부 전체에서 맞춤형 테라조 바닥재는 대부분 개방형으로 설계된 거실 층 내의 공간을 정의하는 반면, 전체 길이의 맞춤형 화이트 오크 목공 작업은 주방뿐 아니라 엔터테인먼트 및 보관 공간도 숨겨준다. 재료 사용의 축소를 통해 벽돌 베일에 대한 집중도를 높이고 조용하고 미니멀한 내부가 휴식을 돕는다. 베일 하우스는 정신없는 거리에서의 삶에서 벗어나 도시 외관 이면에 아늑하고 바람이 잘 통하는 라이프스타일을 구축하려는 열망이 깃들어있다.

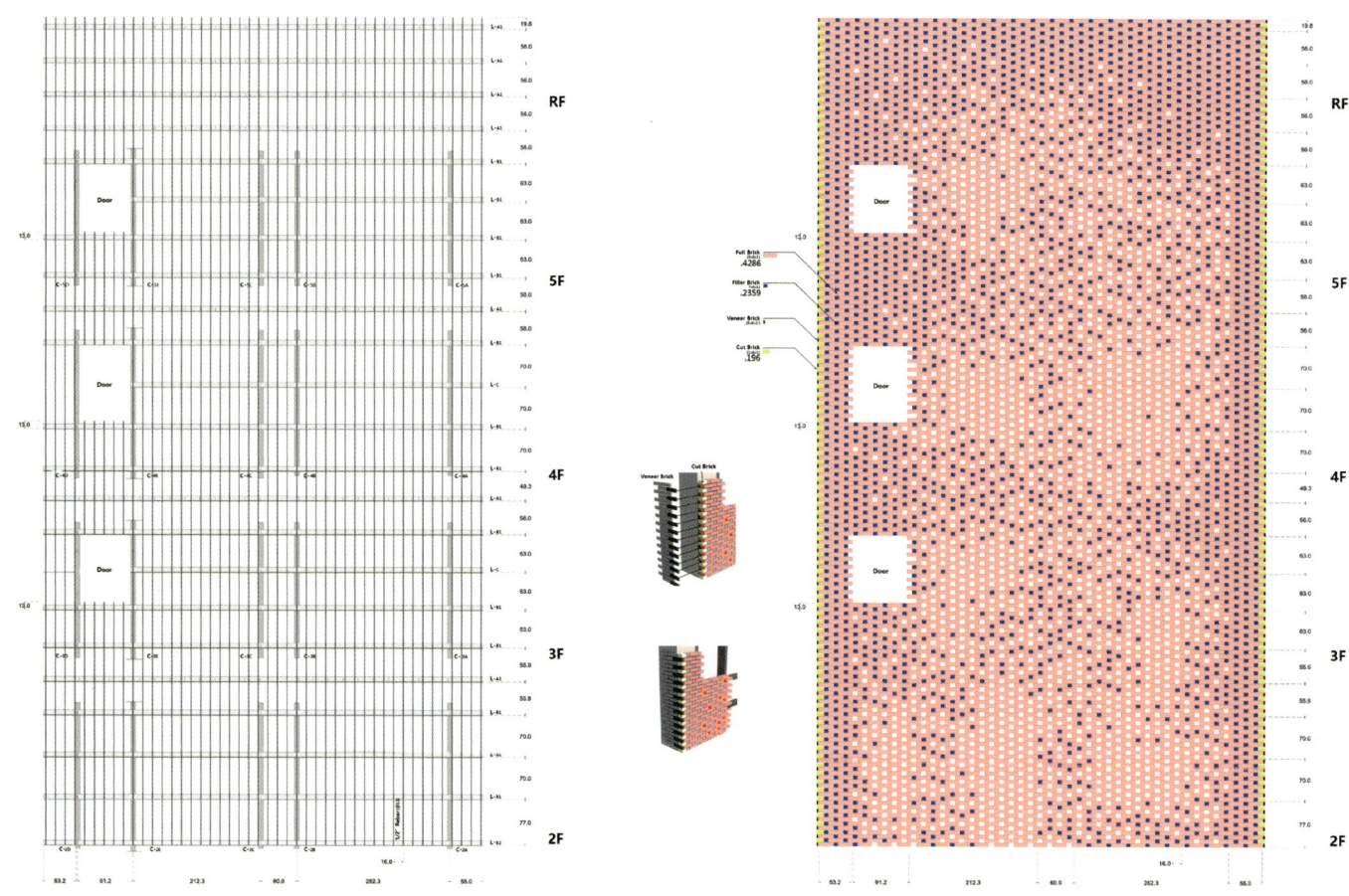

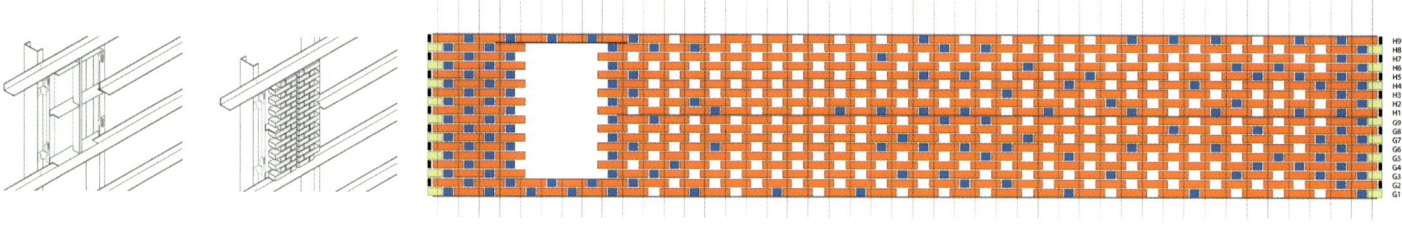

BRICK SURFACE + LAYOUT

← Detail of the brick facade ↱ Detail of the brick facade ↳ Detail of the brick facade

← Entrance　→ Brick facade view from indoor

1　EMERGENCY EXIT
2　STEEL CHANNEL CH200*80*7.5*11
3　M20 ADHESIVE ANCHOR
4　STEEL CHANNEL CH200*80*7.5*11
5　350mm * 250mm STEEL PLATE
6　1/2' SS THREADED REBAR
7　STEEL SHELF ANGLE L150*90*9T WITH 1/2" A325 BOLTS
8　REINFORCED CONCRETE COLUMNS

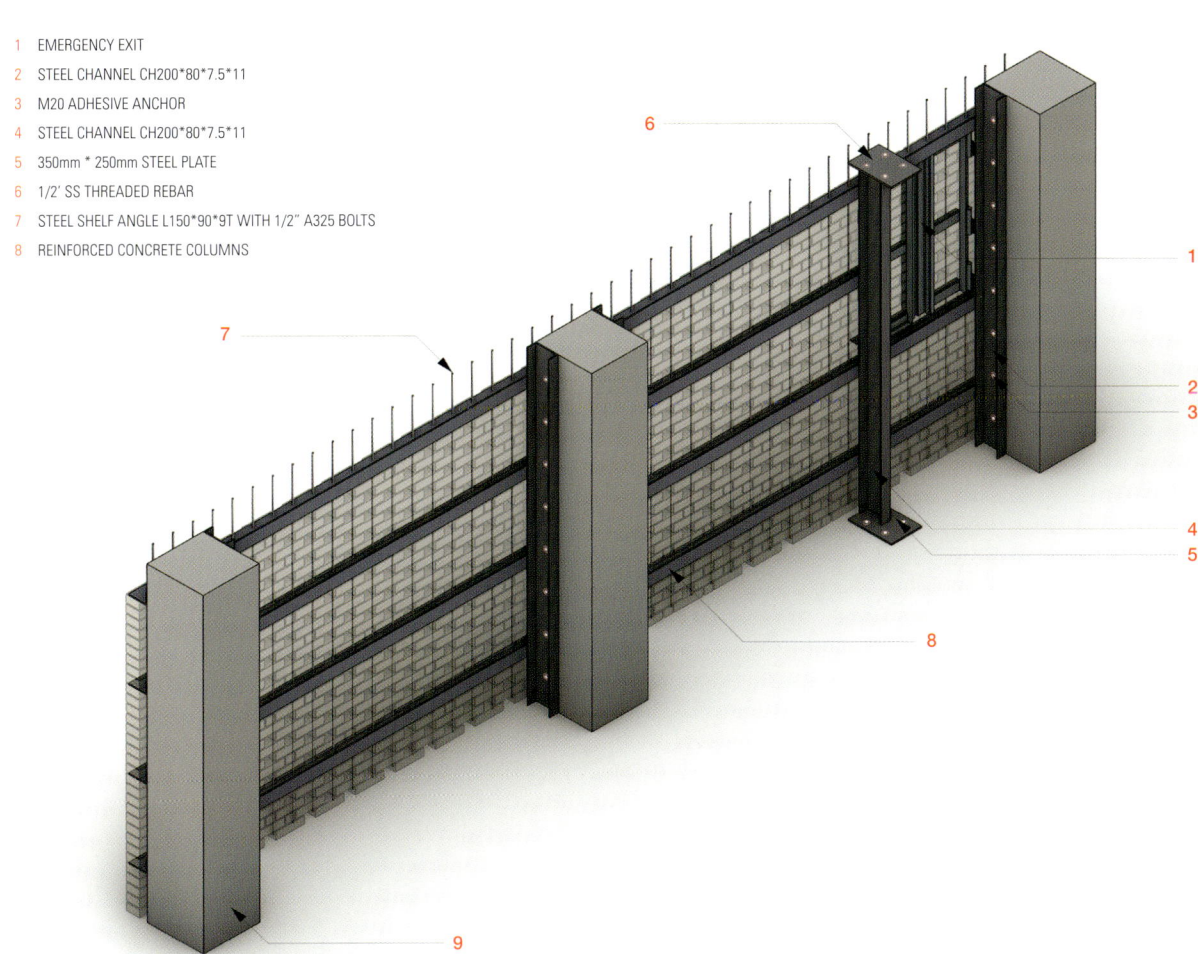

FACADE STRUCTURE DIAGRAM

← Living room → Courtyard

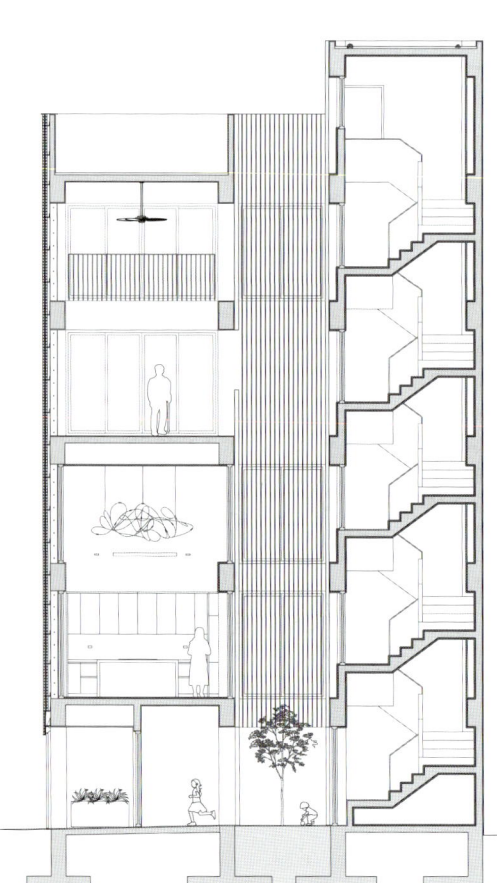

1. 2cm DIA. ROUND HOLES, 16cm O.C.
2. WELD
3. STEEL WIRE MESH TACK WELDED TO PLATE
4. GALVANIZED SHELF ANGLE L150*150*12
5. 1/2" ADHESIVE ANCHOR, 25cm O.C.
6. 6mm GALVANIZED STIFFENERS 48cm O.C.
7. 6mm GALVANIZED STEEL PLATE
8. FULL BRICK
9. 1/2" REBAR, 16cm O.C.
10. VENEER BRICKS
11. GALVANIZED SHELF ANGLE L150*90*8, BOLTED TO STEEL CHANNEL
12. GALANIZED STEEL CHANNEL CH200*80*7.5*11, ANCHORED TO R.C. COLUMN
13. FLASHING
14. GARAGE MOTOR, TRACK AND DOOR

SECTION

SECTION DETAIL

↱ Dining room
→ Kitchen
↳ Kitchen & Dining room

Loggia

← Bedroom → Bedroom

1	ENTRANCE	5	FOYER	9	TOILET	13	LOGGIA
2	URBAN GARDEN	6	LIVING ROOM	10	BEDROOM	14	PRIMARY BEDROOM
3	PARKING LOT	7	KITCHEN	11	KID'S ROOM	15	DRESS ROOM
4	COURTYARD	8	DINING ROOM	12	BATHROOM	16	ROOFTOP TERRACE

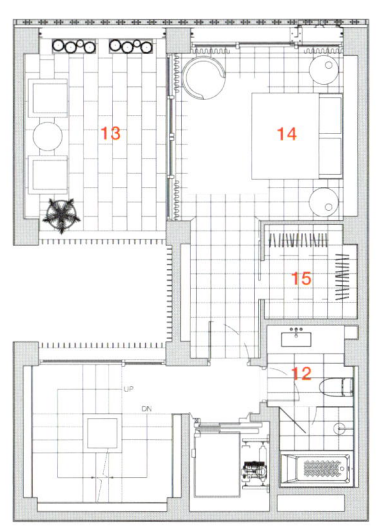

4TH FLOOR PLAN

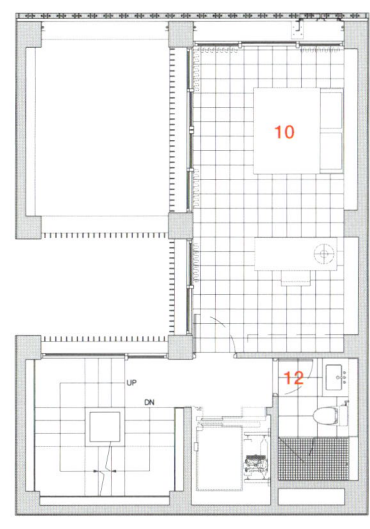

5TH FLOOR PLAN

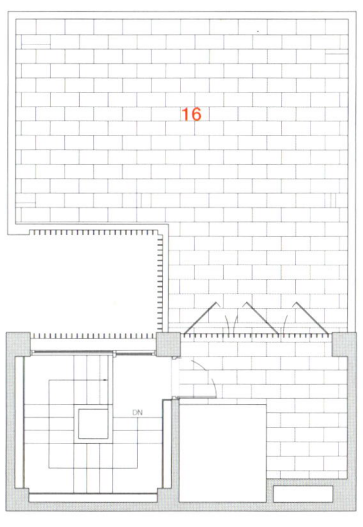

ROOFTOP FLOOR PLAN

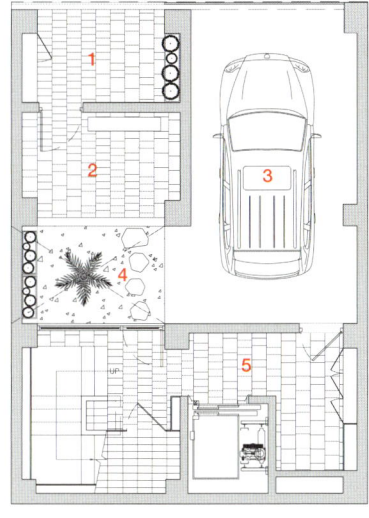

1ST FLOOR PLAN

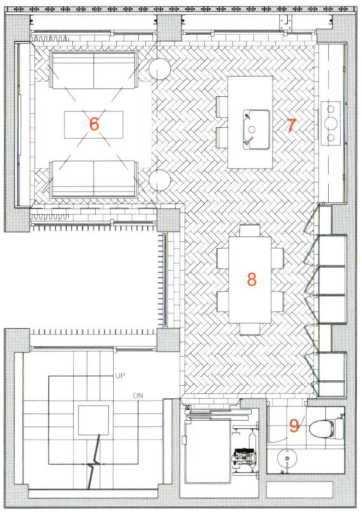

2ND FLOOR PLAN

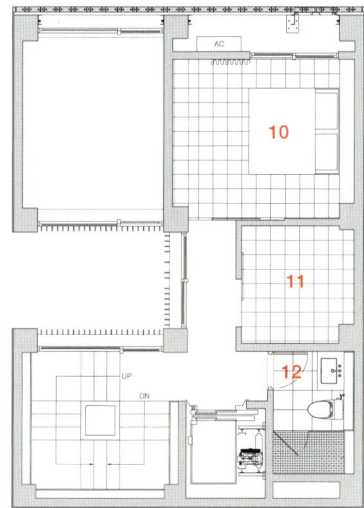

3RD FLOOR PLAN

노보리 빌딩

NOBORI BUILDING

ARCHITECT : FLORIAN BUSCH ARCHITECTS / FLORIAN BUSCH

AT THE TOP OF THE SLOPE LEADING TOWARDS TOKYO'S NUKEBENTEN, the unpretentious appearance of Itsukushima shrine suggests little of its history. In 1086, traversing the Kanto plains on the way to Mutsu, Minamoto no Yoshiie chose this location as temporary camp for his army. The highest in the area, it was easy to spot any approaching danger — and had great views of Mt. Fuji. A year later, returning victoriously, Yoshiie laid the foundation for the shrine which still exists today. The actual site for this project is a 48m² small trapezoid next to the shrine.

EXTRUSION : When your maximum footprint is 80% of 48m² and an ambitious brief calls for a restaurant and several apartments above, the strategy tends to be straightforward: extrude this maximum footprint as far up as possible. Any freedom to explore what is possible must be found in vertical generosity.

CLIMB : As if the client had anticipated the challenges, Nobori, the name given to the project long before there even was a site, means to rise, to climb. The question is —pleasantly pragmatic— how? Where convention would suggest a compact core for circulation, we propose a counter-intuitive move: Dissolve the core to let the stairs climb up and down around the periphery. Entering the building from the street is like turning into one of Tokyo's myriad back alleys, which are de facto residential lobbies mediating between the scales of the city.

DISCOVERIES : What looks a simple volume from the outside is complex inside. As the stair-path pierces the levels at different points, no floor is alike. The vertical generosity pays off: There is enough height for interstices (between stair-volume and slabs) to afford unexpected usages — and thus augment the usual. The complexity is akin to that of the tearoom, where deliberate details open up the confines of a spatial minimum to a cosmos of discoveries.

SKIN : This spatial ambiguity and polyvalence is revealed on the facade. The scale of the building makes it feasible, even advantageous to let the external skin carry the structural loads. Starting with the maximum potential openings, the degree of porosity evolves over time, responding flexibly to requirements and desires during the design process. The facade is the playing field where multiple constraints (structure, views, light, ventilation, budget…) compete with each other, negotiating an optimum compromise.

VIEWS : What was then, in 1086, in the middle of nature is now in the middle of Tokyo. Views of Mt. Fuji have been long been blocked by ever-growing urban masses. But where roads cut straight through these masses, view corridors reveal Tokyo's vastness. Nine centuries after Yoshiie's eyes scanned the surroundings, nothing here is the same anymore. Yet as we climb the building and collect frames of the city, the strategic advantage of the topography can still be felt today.

Location 8-4 Yochomachi, Shinjuku City, Tokyo **Use** Residential, Restaurant & Bar **Site area** 48.45m² **Built area** 36.26m² **Gross floor area** 185.19m² **Completion** 2023 **Project manager** Dyro Yamashita **Design team** Florian Busch, Sachiko Miyazaki, Mayo Shigemura, Dyro Yamashita, Sie-Jhih Chen, Jeng Pheera, Christian Baumgarten, Joachim Nijs, Reo Shima, Bo-Hao Liu, Kana Takagi (Intern) **Contractor** Shin Corporation **Photographer** Vincent Hecht

← Exterior view → Site view

← Small trapezoid site　→ Exterior view

도쿄의 누케벤텐으로 이어지는 경사면 꼭대기에 있는 이츠쿠시마 신사는 그 겉모습이 그 역사의 중요성을 전혀 암시하지 않는다. 1086년, 무츠로 가던 길에 간토 평원을 가로질러 갔던 미나모토노 요시이에는 자신의 군대를 임시로 주둔시킬 위치로 이곳을 선택했다. 이 지역에서 가장 높은 곳이었기에 쉽게 다가오는 위험을 감지할 수 있었고, 후지산의 훌륭한 전망도 자랑했다. 1년 후에 승리를 거두고 돌아온 요시이에는 오늘날까지도 남아 있는 신사의 기초를 놓았다. 이 프로젝트의 실제 부지는 신사 옆에 있는 48㎡의 작은 사다리꼴 부지이다.

압출: 최대 면적이 48㎡의 80%일 때, 그리고 야심 찬 기획안이 레스토랑과 그 위의 몇 개의 아파트를 요구할 때, 전략은 단순명료하다. 가능한 한 높이가 이 최대 면적을 최대한 멀리 돌출시키는 것이다. 가능한 것을 탐색할 자유는 수직적 넉넉함에서 찾아야 한다.

상승: 고객이 마치 도전과제들을 예상이라도 한 듯, 부지가 확정되기 훨씬 이전에 프로젝트에 붙여진 이름 '노보리'는 상승하다, 오르다,는 뜻을 가지고 있다. 문제는 -유쾌하게도 실용적으로- 어떻게 오를 것인가이다. 일반적인 관례가 소통의 중앙 핵심부를 압축하라고 제안할 때, 우리는 직관에 반하는 행동을 제안한다. 중앙 핵심부를 해체하여 계단이 주변을 따라 오르내릴 수 있도록 코어를 녹이는 것이다. 거리에서 건물 안으로 들어서는 것은 마치 도쿄의 수많은 뒷골목 중 하나로 들어서는 것과 같고, 이는 사실상 도시의 규모 사이를 중재하는 주거용 로비 역할을 한다.

발견: 외부에서 보는 단순한 볼륨처럼 보이는 것은 내부에 복잡성이 숨어 있다. 계단 경로가 각기 다른 지점에서 층을 관통함에 따라 어떤 층도 서로 동일하지 않다. 세로로 제공된 넉넉함은 그 가치를 발휘한다. 틈새(계단 체적과 슬래브 사이)에 충분한 높이를 확보함으로써, 예상치 못한 용도의 공간을 제공하고 따라서 범한 것을 향상시킨다. 이는 다도실의 복잡성과 유사하며, 고의적인 세부 사항을 통해 최소한의 공간적 한계를 발견의 우주로 열어준다.

스킨: 이 공간의 모호함과 다목적성은 파사드에 드러난다. 건물의 규모는 외부 스킨 구조적 부하를 지탱하게 하는 것을 가능할 뿐만 아니라 유리하게 한다. 최대한의 개구부부터 시작하여, 투과성의 정도는 디자인 과정 동안 요구와 바람에 유연하게 반응하면서 시간에 따라 진화한다. 파사드는 다양한 제약 조건(구조, 전망, 빛, 통풍, 예산 등)이 서로 경쟁하고 최적의 타협을 협상하는 장이다.

전망: 1086년 당시 자연 한가운데에 있던 것이 이제는 도쿄 한복판에 자리 있다. 후지산의 전망은 계속해서 성장하는 도시의 덩어리들에 의해 오래전에 가려졌다. 그러나 도로가 이 덩어리들을 곧장 관통할 때, 전망 통로는 도쿄의 광활함을 드러낸다. 요시이에가 주변을 살피던 지난 9세기 후, 이곳의 모든것은 더 이상 예전과 똑같지 않다. 그러나 건물을 오르며 도시의 틀을 수집할 때 지형의 전략적 이점은 오늘날에도 여전히 느껴진다.

↑ Street view

ELEVATION

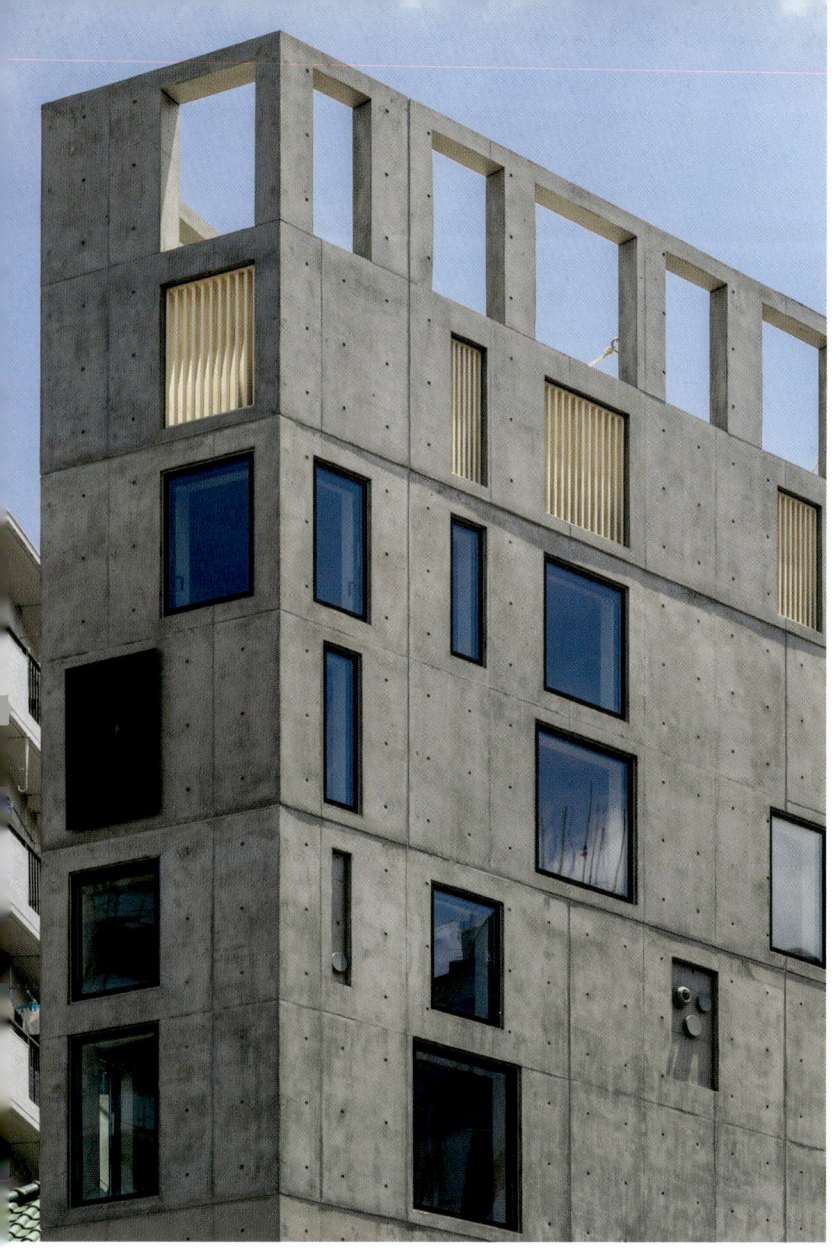

← Facade & window view　→ Facade & window view　↲ Facade & window view　↳ Restaurant entrance

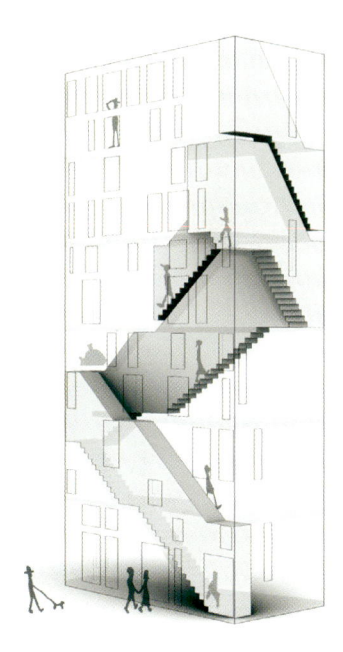

SKETCH

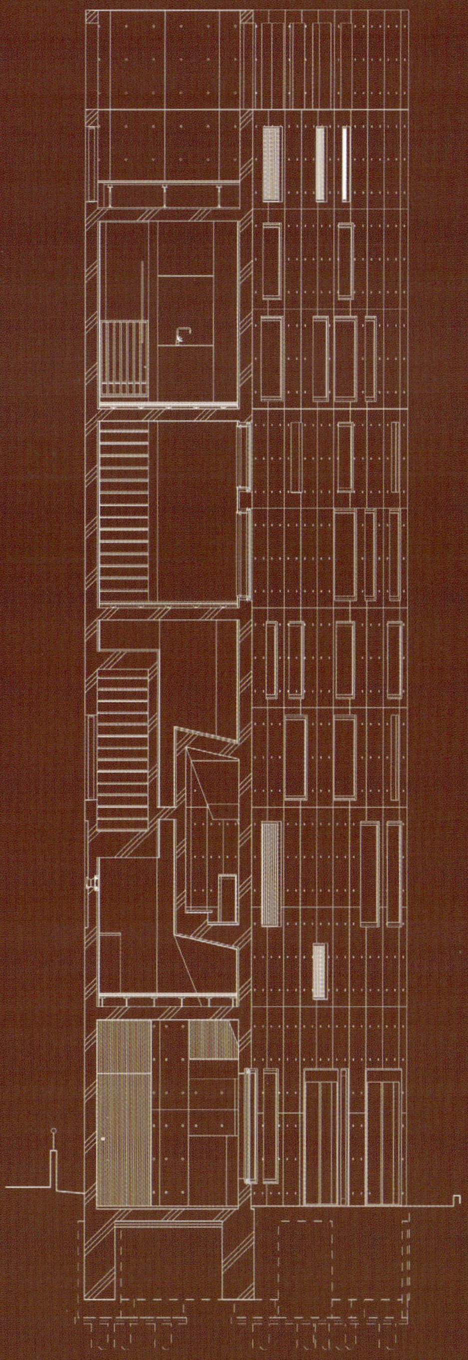

SECTION

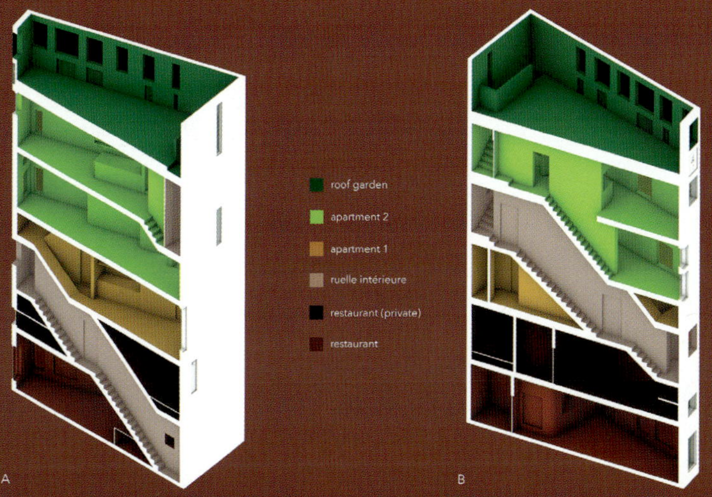

SECTION

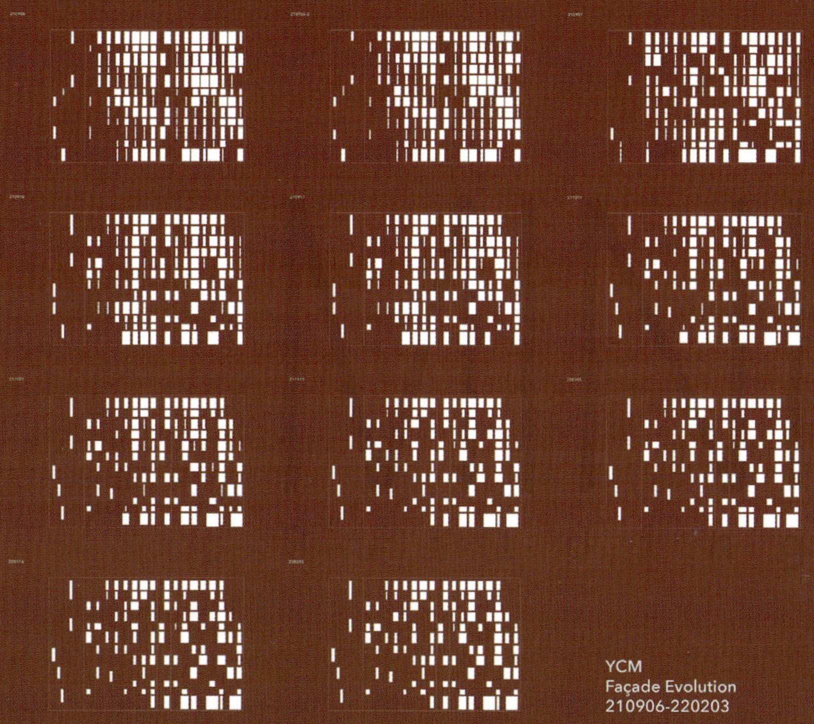

FACADE EVOLUTION

YCM
Façade Evolution
210906-220203

MODELING & POROSITY STUDIES

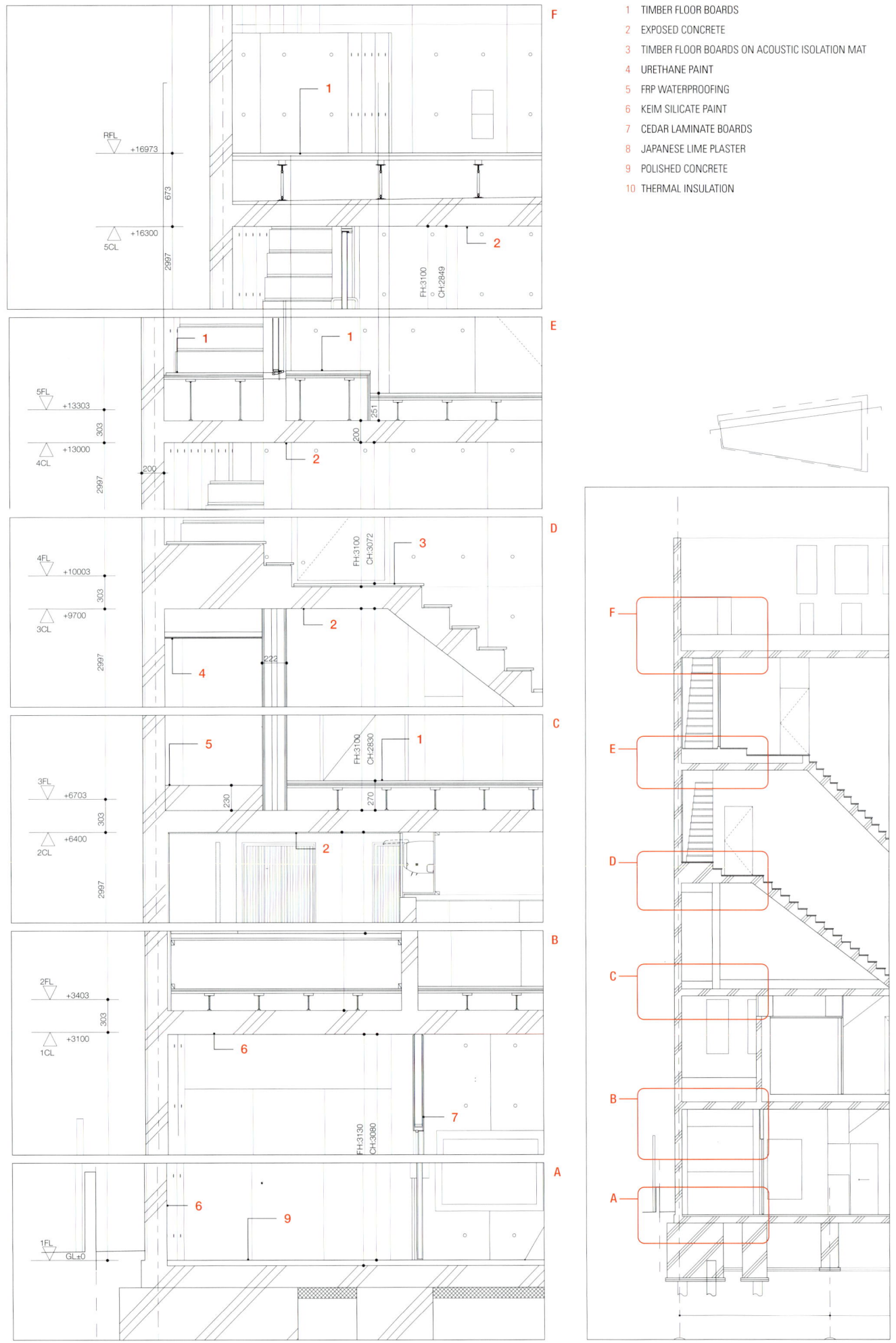

1	TIMBER FLOOR BOARDS
2	EXPOSED CONCRETE
3	TIMBER FLOOR BOARDS ON ACOUSTIC ISOLATION MAT
4	URETHANE PAINT
5	FRP WATERPROOFING
6	KEIM SILICATE PAINT
7	CEDAR LAMINATE BOARDS
8	JAPANESE LIME PLASTER
9	POLISHED CONCRETE
10	THERMAL INSULATION

SECTION DETAIL

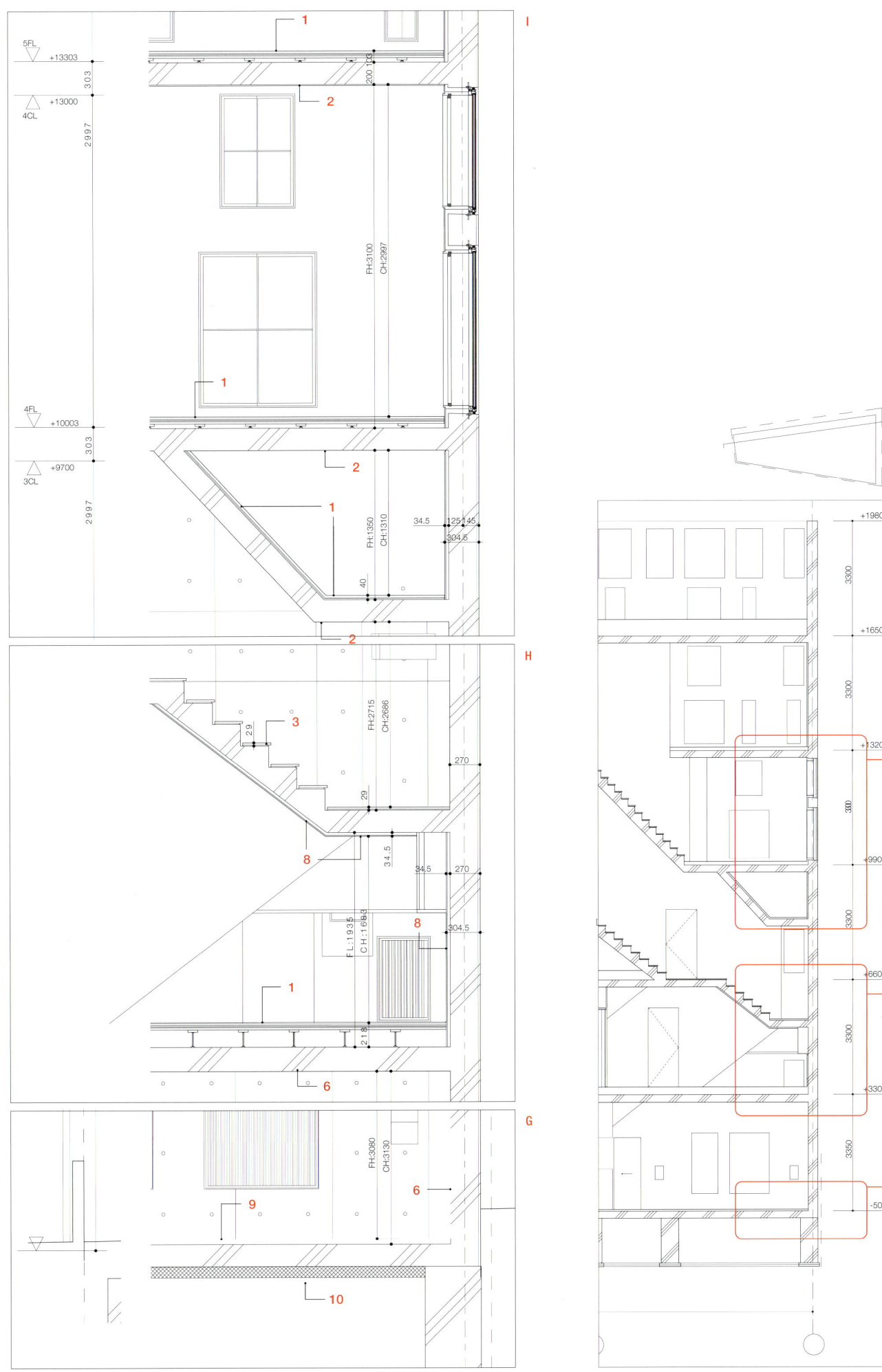

SECTION DETAIL

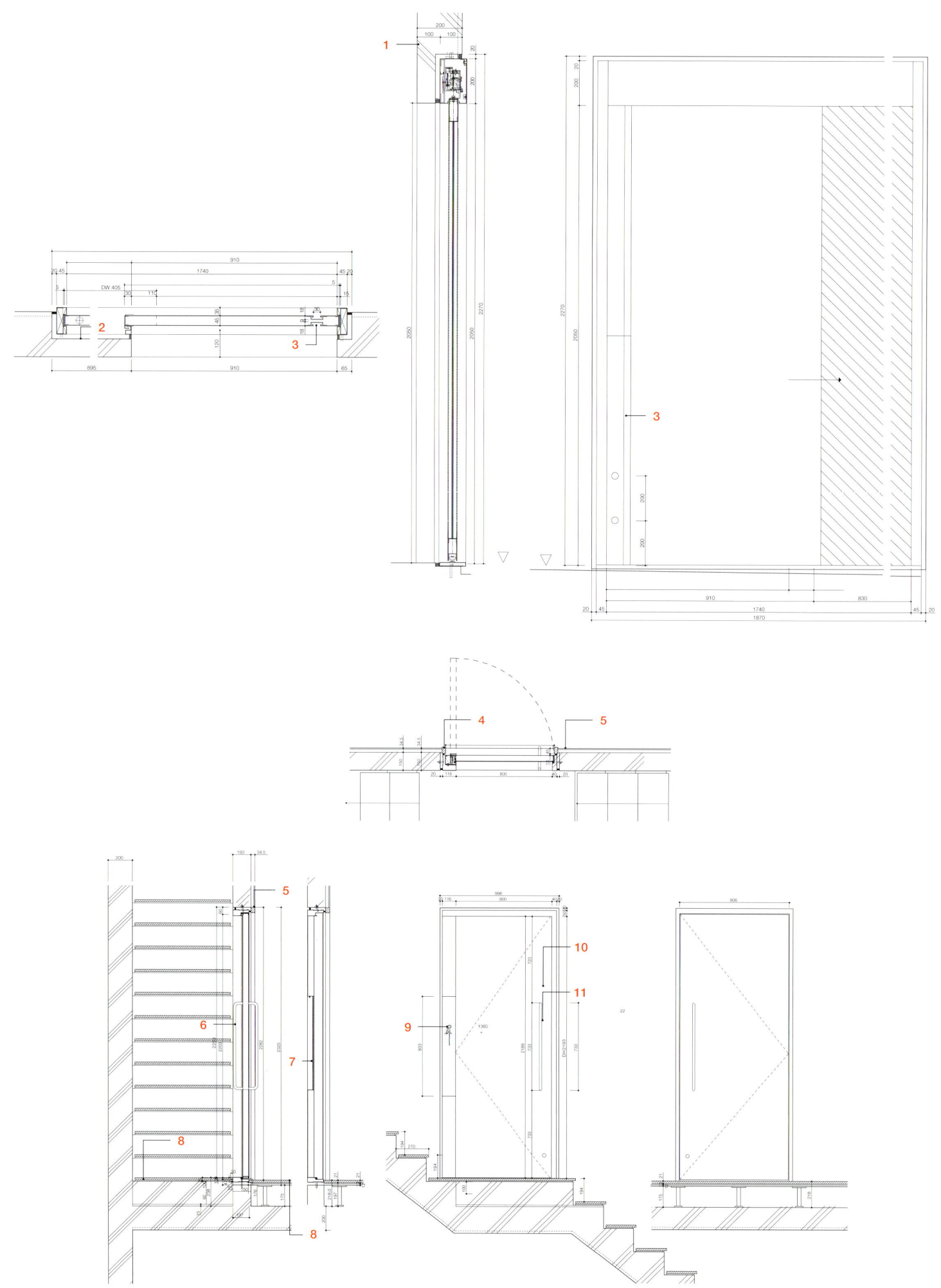

DETAILS

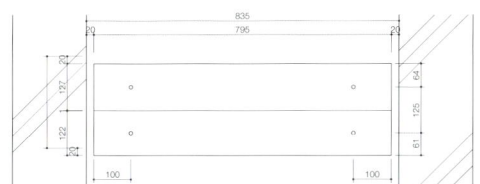

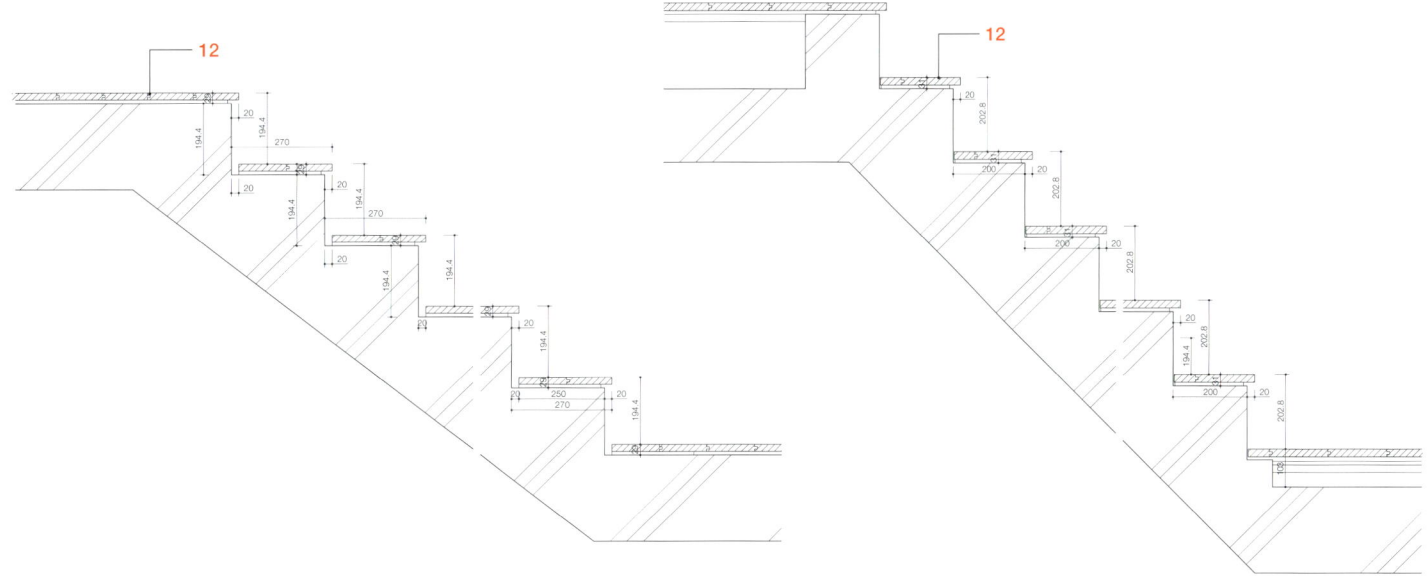

DETAILS

1. EXPOSED CONCRETE
2. BLACK STEEL PLATE
3. DOOR HANDLE SLIT
4. ACRYLIC EMULSION PAINT
5. JAPANESE LIME PLASTER
6. BLACK STEEL PIPE
7. ETCHED CEDAR SIGN
8. TIMBER FLOOR BOARDS
9. BLACK STEEL SIGN (UNTREATED)
10. BLACK STEEL (UNTREATED)
11. BLACK STEEL PIPE
12. TIMBER FLOOR BOARDS ON ACOUSTIC ISOLATION MAT

← Stair circulation system window ↑ Stair → Stair & circulation system window

↑┐ Apartment interior ┌↑ Apartment kitchen ← Apartment kitchen → Apartment kitchen ↓┘ Restaurant interior └↓ Restaurant interior

← Rooftop   ↑ Staircase   → Toilet

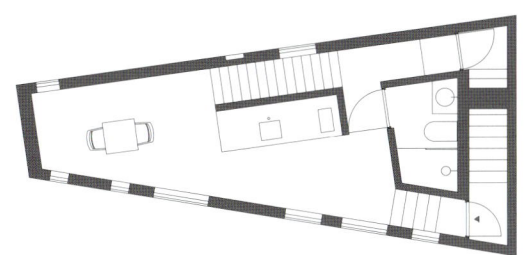

5TH FLOOR PLAN_APARTMENT

ROOFTOP PLAN

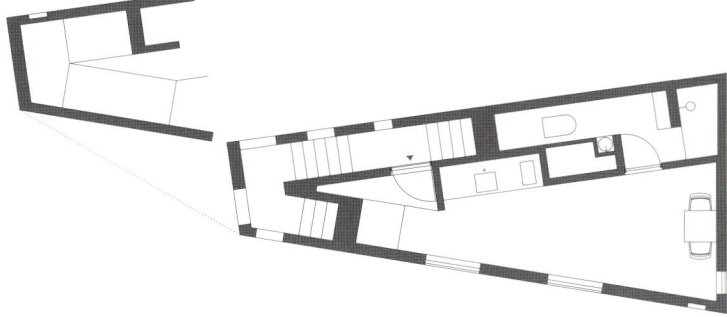

3RD FLOOR PLAN_APARTMENT

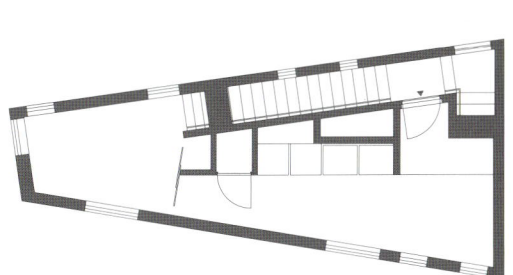

4TH FLOOR PLAN_APARTMENT

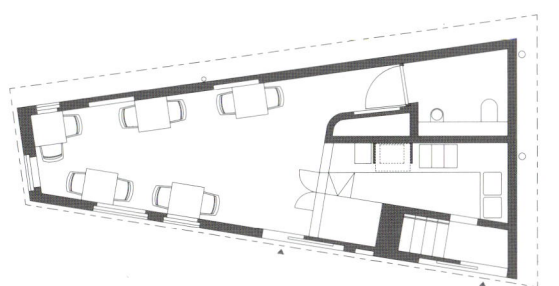

1ST FLOOR PLAN_RESTAURANT

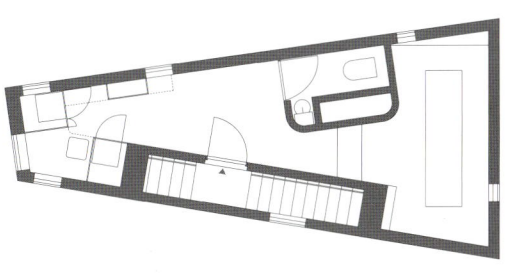

2ND FLOOR PLAN_RESTAURANT

태국 모던 알루미늄 박물관

MUSEUM OF MODERN ALUMINUM THAILAND

ARCHITECT : HAS DESIGN AND RESEARCH / JENCHIEH HUNG, KULTHIDA SONGKITTIPAKDEE

MOMA IS THE ABBREVIATION FOR MUSEUM OF MODERN ALUMINUM Thailand. The project originated from a group of ambitious clients with the goal of reviving the significance of aluminum in Thailand.

MoMA is located at the busiest traffic hub on the outskirts of Bangkok, where heavy traffic has led to a variety of commercial signs lining Ratchaphruek Road. The main roads lead to The Grand Palace, Wongwian Yai, Bangkok University, and Ko Kret, the only island in Bangkok. More than a decade ago, fireflies populated Ko Kret, making the island a natural retreat for Bangkokians.

HAS Design and Research wanted MoMA to serve not only as a public space but also as a getaway for busy urban dwellers. The building extends the natural landscape of Ko Kret Island to the project site. During the day, MoMA is a dandelion, with its overhanging elements swaying in the wind, bringing softness and lightness to the busy Ratchaphruek Road; at night, MoMA transforms into a firefly, adding a sense of nature and peacefulness to the highly commercialized Ratchaphruek Road.

MoMA not only uses aluminum strips as display items, but also allows them to continue in the architecture, the interior, the landscape, as well as the lighting and furniture, creating a sense of totality inside and outside. The façade is clad with tens of thousands of aluminum strips, each with a slightly different color and texture, just like the feathers of a dandelion. The aluminum strips, combined with LED lighting, extend from the front facade to the two side façades, and then straight through the "tunnel" space on the west side, filtering and dampening the noise of the external environment and guiding visitors to the quiet exhibition place.

The aluminum strips on the facade not only provide a variety of lighting functions, but also shade the interior from excessive sunlight to maintain a comfortable interior environment. The flexibility of the exhibition space can meet a variety of display, reception, and activity needs. On the top floor, the enclosed landscape resembles a floating island with seasonal plants, creating an urban ecological site for fireflies to flourish.

Location Nonthaburi, Thailand **Use** Gallery, Exhibition, Retail, Office **Site area** 250 m² **Built area** 180m² **Gross Floor area** 400m² **Completion** 2022 **Project principal** Jenchieh Hung, Kulthida Songkittipakdee **Design team** Jenchieh Hung, Kulthida Songkittipakdee, Jiaqi Han, Qinye Chen **Contractor** SL Window Co., Ltd. **Landscape design** TROP : terrains + open space **Lighting design** Light Is **Facade engineer** AB&W Innovation Co., Ltd. **Lighting engineer** Neowave Technology **Photographer** W Workspace

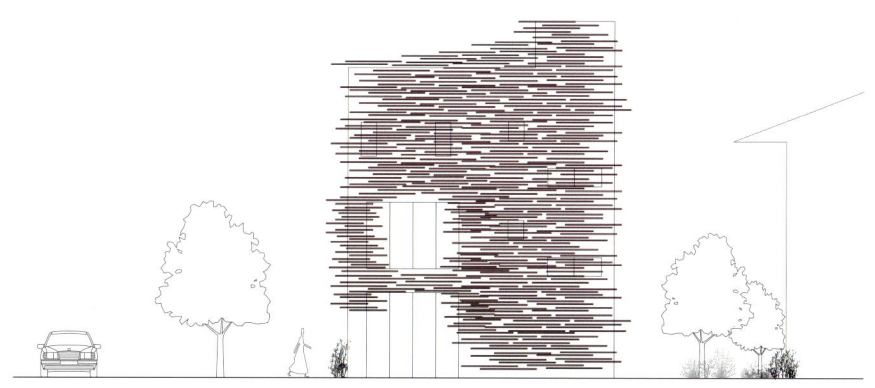

SKETCH

← Exterior night north view

MoMA는 Museum of Modern Aluminium Thai의 약자입니다. 이 프로젝트는 태국에서 알루미늄의 중요성을 되살리고자 하는 야심 찬 클라이언트들로부터 시작되었다.

MoMA는 방콕 외곽의 가장 번잡한 교통 허브에 있으며, 밀집한 교통으로 인해 라차프루엑 도로에는 다양한 상업 간판들이 있다. 주요 도로들은 그랜드 팰리스, 웡위안 야이, 방콕 대학교, 그리고 방콕에서 유일한 섬인 코 크레트로 이어진다. 십여 년 전, 반딧불이가 코 크레트 섬을 가득 메우며 방콕 시민들에게 자연의 피난처를 제공했다.

HAS 디자인 앤 리서치는 MoMA가 단순한 공공 공간을 넘어 바쁜 도시민들에게 일상에서 벗어날 수 있는 휴식처가 되기를 원했다. 이 건물은 코 크레트 섬의 자연 풍경을 프로젝트 현장까지 확장했다. 낮에는 라차프루엑 도로의 분주한 속에서 부드러움과 가벼움을 더하는 풍경처럼, MoMA는 바람에 흔들리는 민들레와 같고 밤에는 반딧불로 변신하여 상업화된 라차프루엑 도로에 자연과 평화로움을 더해 준다.

MoMA는 알루미늄 스트립을 전시 아이템으로만 사용뿐만 아니라, 건축물은 물론 내부 공간, 조경, 그리고 조명과 가구에 이르기까지 알루미늄 스트립을 계속 이어 나가며 내외부에 통일감을 부여하였다. 파사드는 민들레 깃털처럼 각기 다른 색상과 질감을 지닌 수만 개의 알루미늄 스트립으로 마감되었다. LED 조명과 결합한 알루미늄 스트립들은 전면 파사드에서 시작해 양쪽 옆면 파사드를 거쳐 서측의 "터널" 공간을 직진하면서 외부 환경의 소음을 필터링하고 감소시켜 방문객들을 조용한 전시 공간으로 이끈다.

파사드의 알루미늄 스트립은 다양한 조명 기능을 제공할 뿐만 아니라, 과도한 햇볕으로부터 실내를 가려 쾌적한 내부 환경을 유지한다. 전시 공간의 유연성은 다양한 디스플레이, 리셉션 및 활동 요구사항을 충족할 수 있다. 최상층의 폐쇄된 조경은 계절에 따라 변화하는 식물들이 떠 있는 섬을 연상시켜, 반딧불이 번성할 수 있는 도시 생태 장소를 조성하였다.

↑ Surrounding view

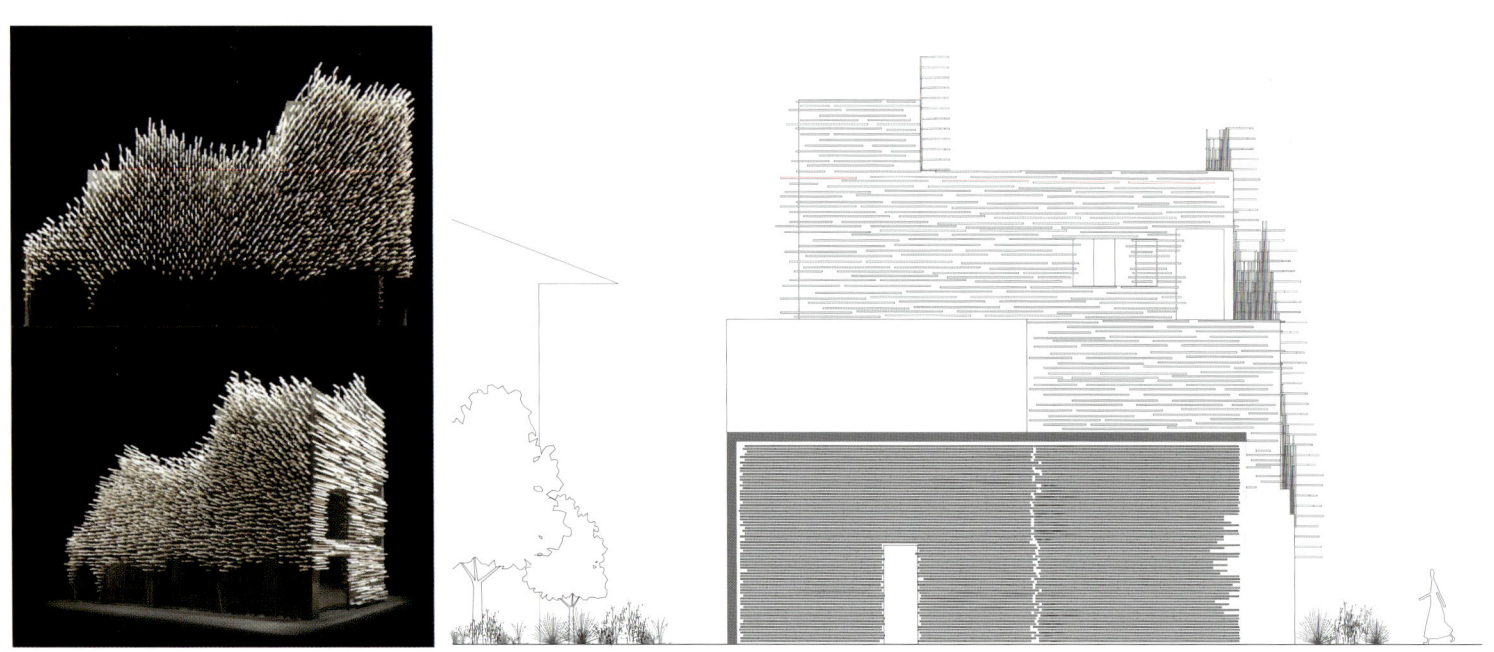

MODELING   SECTION

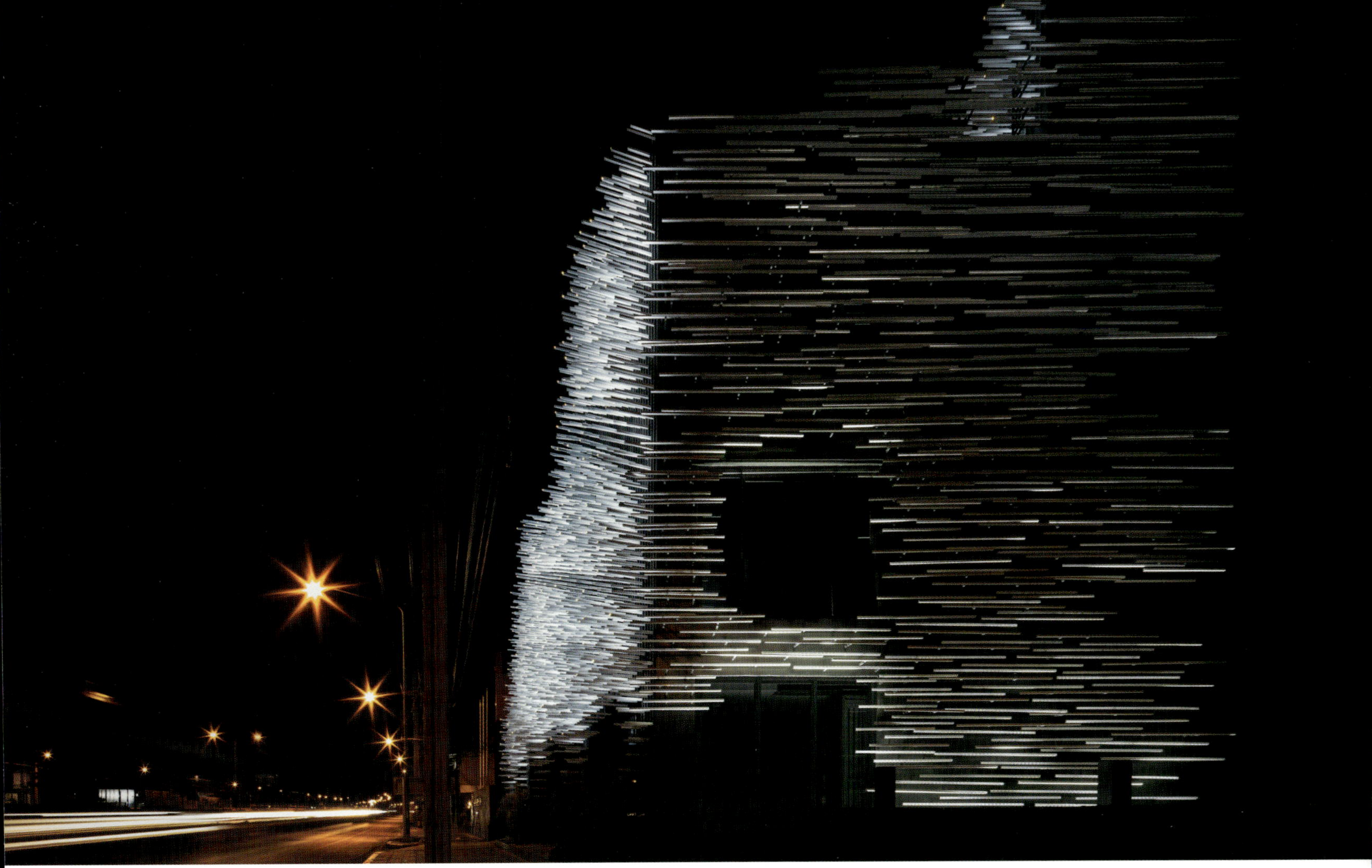

↑ Night image view

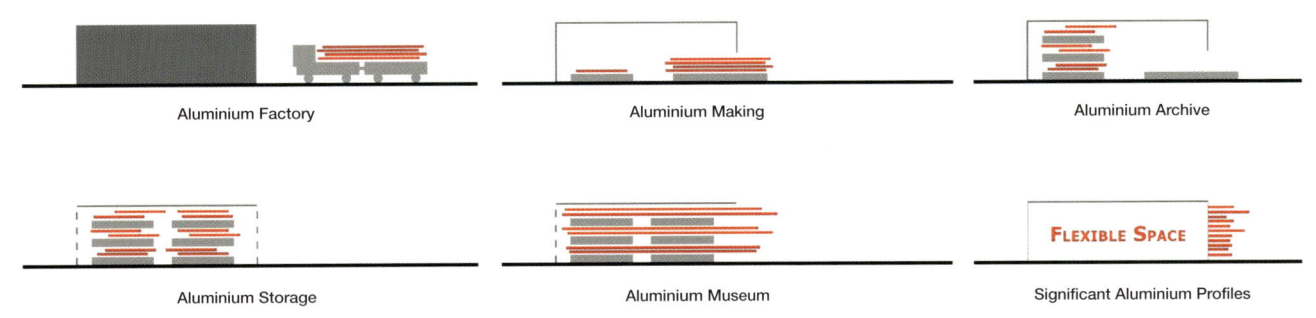

Aluminium Factory
Aluminium Making
Aluminium Archive
Aluminium Storage
Aluminium Museum
FLEXIBLE SPACE
Significant Aluminium Profiles

CONCEPT DIAGRAM

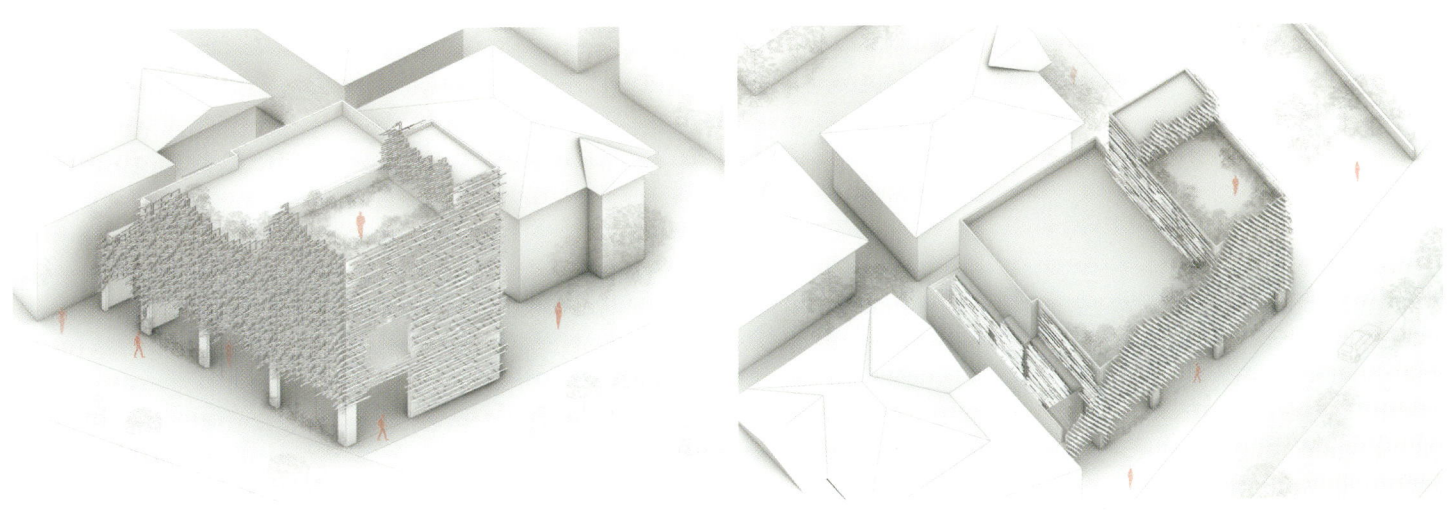

ISOMETRIC DIAGRAM

Thousands of aluminum strips | Entrance

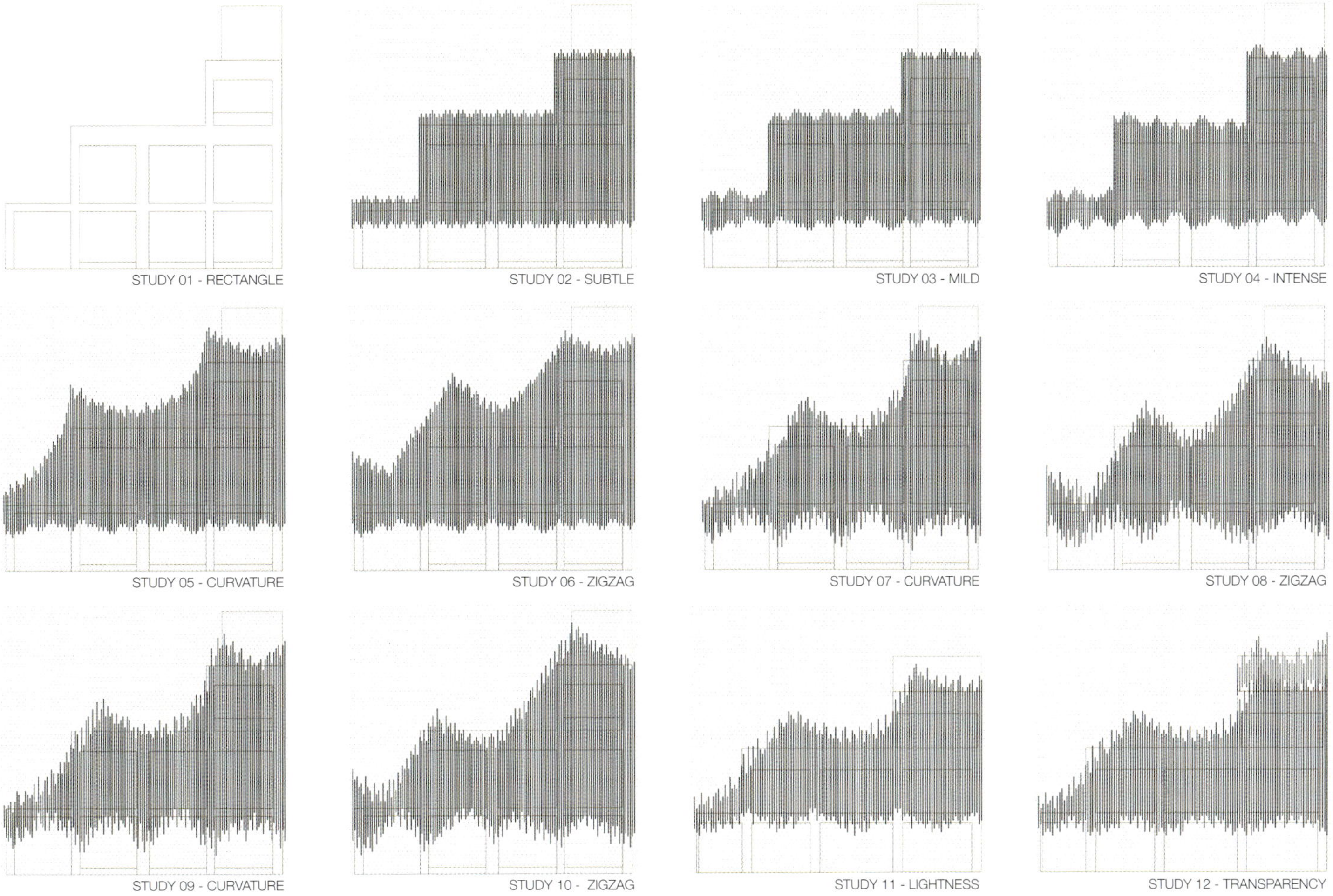

STUDY 01 - RECTANGLE	STUDY 02 - SUBTLE	STUDY 03 - MILD	STUDY 04 - INTENSE
STUDY 05 - CURVATURE	STUDY 06 - ZIGZAG	STUDY 07 - CURVATURE	STUDY 08 - ZIGZAG
STUDY 09 - CURVATURE	STUDY 10 - ZIGZAG	STUDY 11 - LIGHTNESS	STUDY 12 - TRANSPARENCY

FACADE STUDY

LENGTH : 350 MM.
LENGTH : 450 MM.
LENGTH : 550 MM.
LENGTH : 650 MM.

MOCKUP

↑ Roof top garden　← Image of feathers a dandelion　↙ Image of feathers a dandelion　→ Thousands of aluminum strip

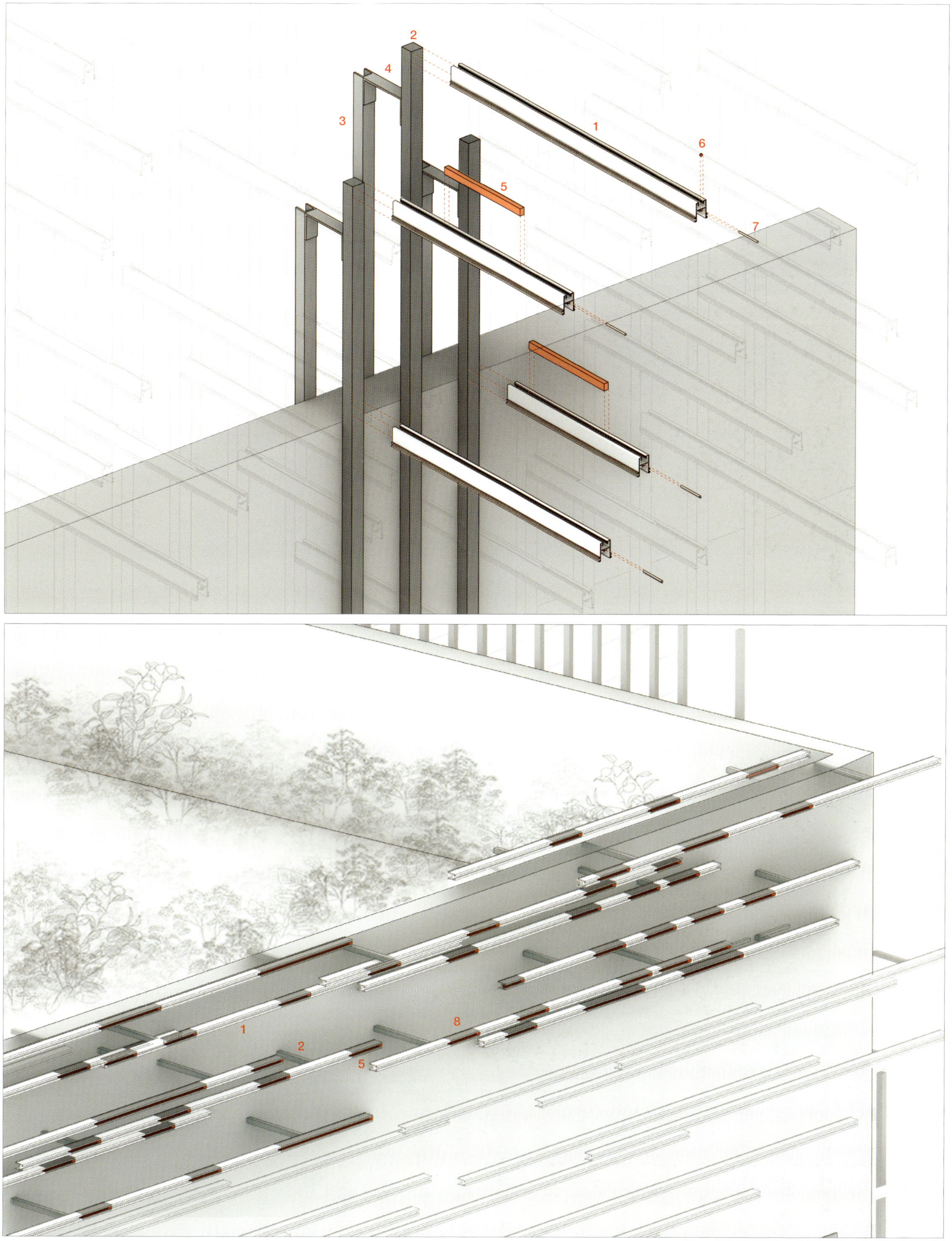

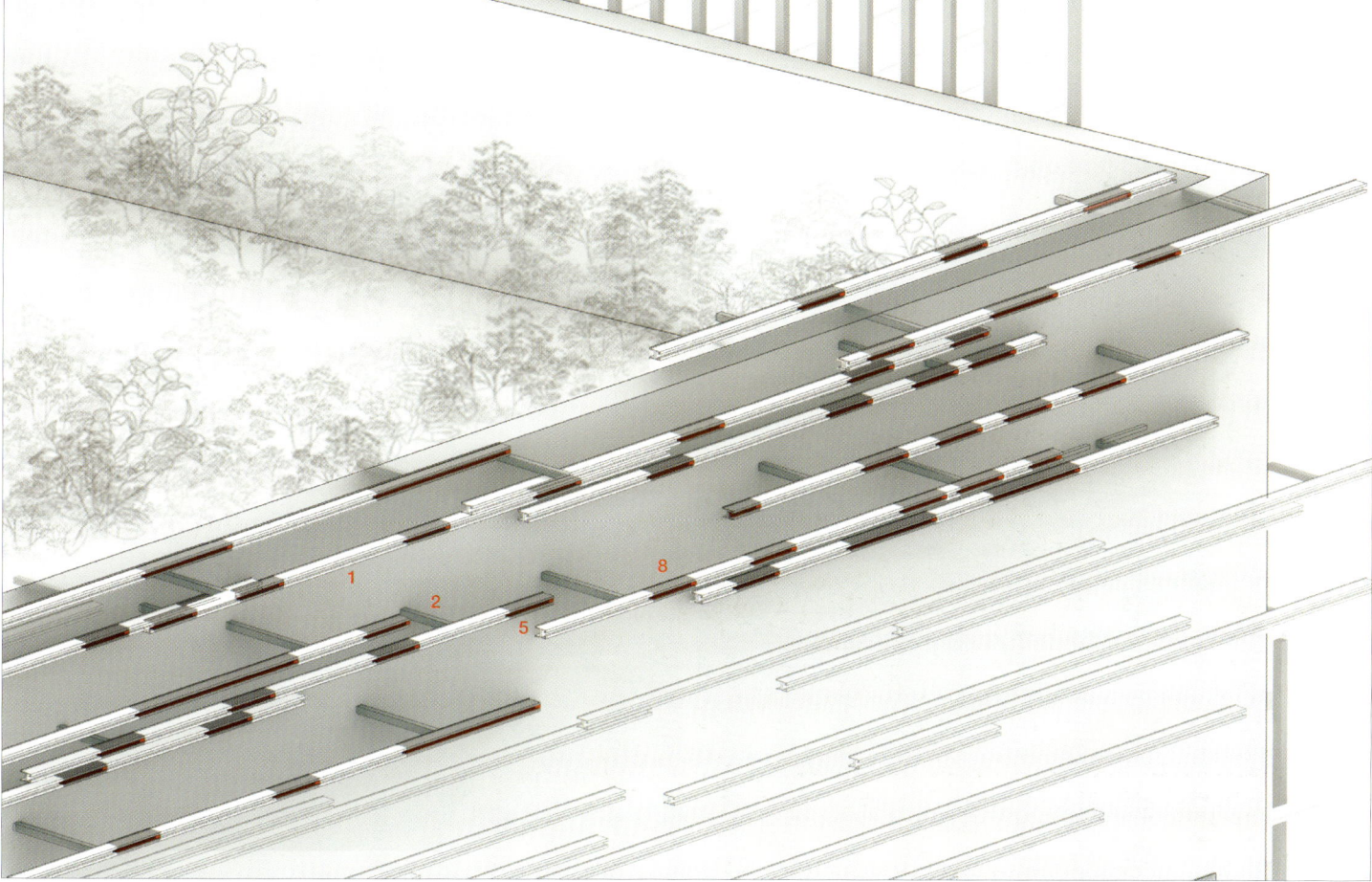

1 ALUMINIUM PROFILES (LENGTH: 350, 450, 550, 650mm)
2 ALUMINIUM STRUCTURE (25X25mm)
3 C STEEL STRUCTURE
4 C STEEL BRACKET
5 LED STRIP LIGHT
6 LED DOT LIGHT
7 HIDDEN SCREW
8 CHARCOAL GRAY PAINT

ALUMINUM STRIPS LIGHTING DETAIL

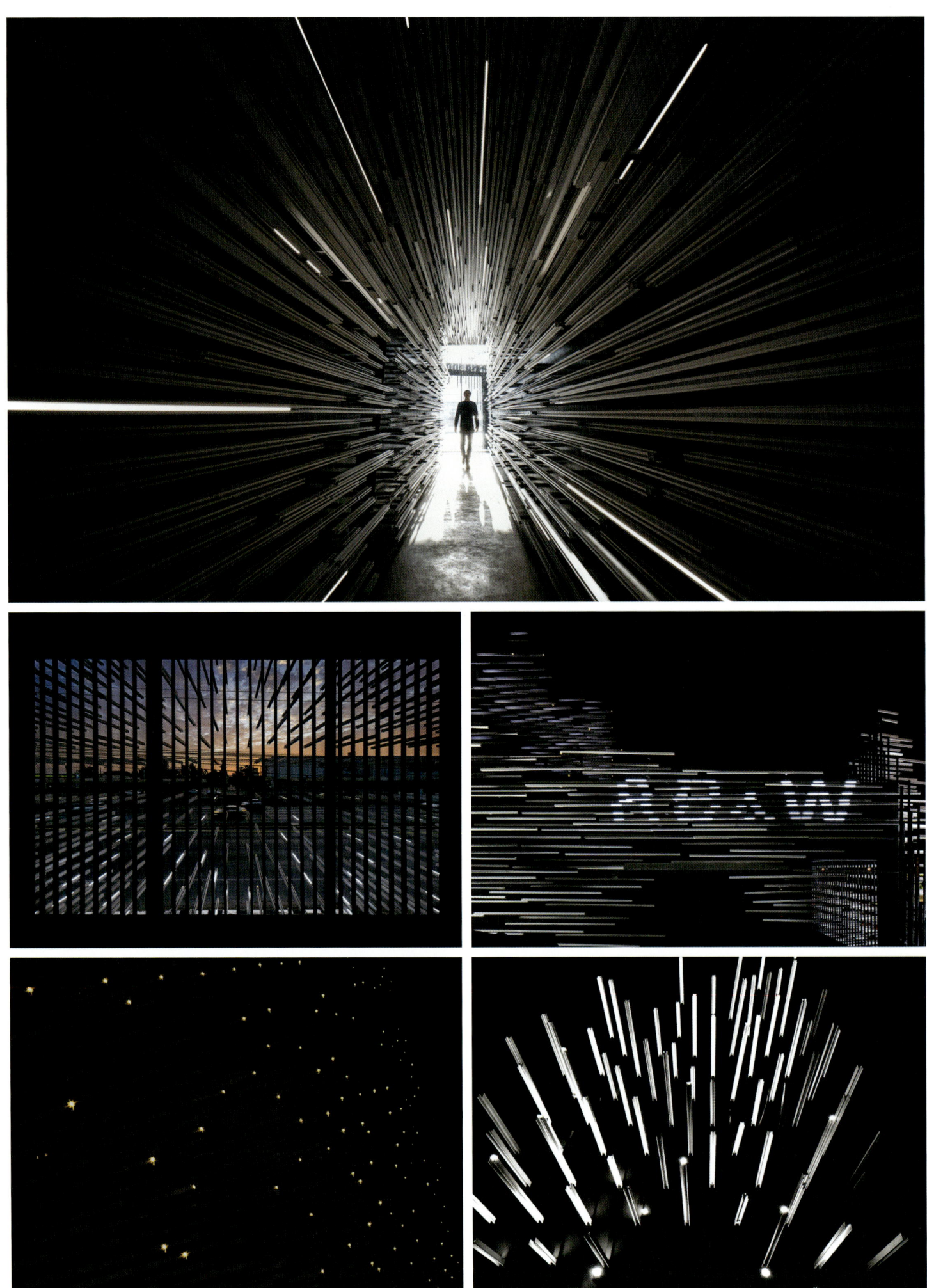

↑ 1st floor entrance & hallway   ← Outside view through the window   → Sign lighting design   ↙ Lighting design   ↘ Lighting detail view

↑ Various aluminum strip view

SITE PLAN

프렌치 키치 3 카페
FRENCH KITSCH Ⅲ CAFE

ARCHITECT : TOUCH ARCHITECT / SETTHAKARN YANGDERM, PARPIS LEELANIRAMOL

THE MAIN DESIGN CONCEPT OF FRENCH KITSCH IS DEVELOPED by interpreting its name; "French" and "Kitsch". Apart from being a specialized French patisserie, the owner's love for French bulldog also comes as brand identity, portraying a playful yet elegant image to the cafe. The design aims to enhance this image together with the concept of "Kitsch", a form of art that appreciates imperfection through architecture elements and materials.

The French cathedral is taken as a primary reference to the design where rhythmic arches are developed. Instead of symmetrical arches, imperfect arches of different scales are used. The design started from a perfect rectangular mass which is made imperfect by carving out imperfect arches on the first level and inverted imperfect arches on the second level. On the 1st floor, these arches embrace visitors with their antique yet modern looks, creating shadow along the path and when light passes through the arched window, it creates reflection on the floor, similar to that of cathedral glass. The oversized imperfect arch also creates a continuous space from the counter to the 2nd floor, highlighting the full-function counter, allowing it to be seen from both floors. On the 2nd floor, voids are carved into the inverted curves, allowing sunlight to enter.

By using textured concrete, it strengthens the concept of perfection of imperfection where the wall is not completely smooth, but it reflects the authenticity of the material which can be beautiful by itself. Moreover, by using concrete as the main material, the furniture, decorations, and LED lights of green and pink are made outstanding, emphasizing the brand's color identity, creating a strong memorable image of the cafe.

Location City Link, Mueang, Nakornratchasima, Thailand **Use** Cafe **Site area** 615.94m² **Gross Floor area** 360m² **Completion** 2023 **Project manager** Setthakarn Yangderm, Parpis Leelaniramol **Design team** Pitchaya Tiyapitsanupaisal, Tanita Panjawongroj, Thanunya Deeprasittikul, Matucha Kanpai **Civil engineer** Chittinat Wongmaneeprateep **M&E Engineer** Yodchai kornsiriwipha, Isarapap Rattanabumrung **Contractor** Samma Construction **Photographer** Metipat Prommomate, Anan Naruphantawat

Exterior view

FINAL DESIGN

This project is designed basically based on pracmatic concern together with the branding name **'French Kitsch'**. This building is not only an architecture, but also act as a sculpture in an ancient art era until modern development, in which the most of architectural elements are used to create space and give new experience due to its definition.

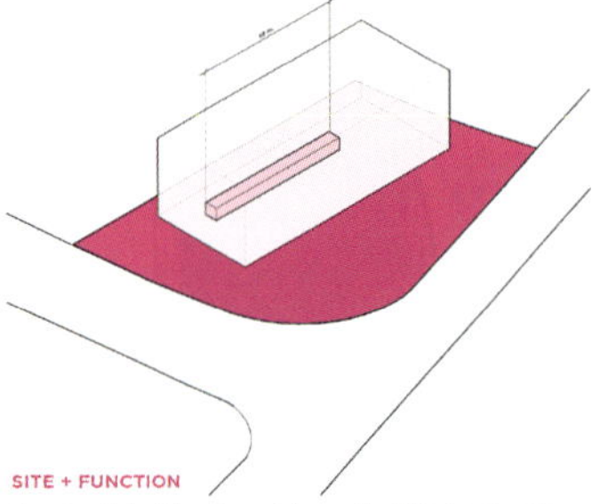

SITE + FUNCTION

A rectangle building shape is formed by following the site shape to reach its fullest potential of area usage, where a minimum of 12-meter-long counter is also required.

CLIMATE CONCERN / ACCESSIBILITY

Due to a long part of building shape faces West, the hottest direction, while needed to be a main entrace since it is a node of accessibility. A special architecture elements is needed to attract people, while preventing heat from direct sunlight.

French kitsch

French cathedral, historical architecture, the most notable in France. Rhythmic arch from these religious place is used for creating the space.

French **kitsch**

Perfection of imperfection, the beauty of **Kitsch** art is applied in the building to reflect the characteristic of the branding.

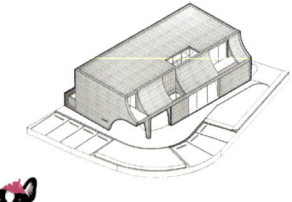

RHYTHMIC ARCH

Rhythm of arch is applied at the highlighted space which creates an antique feeling stunning perception.

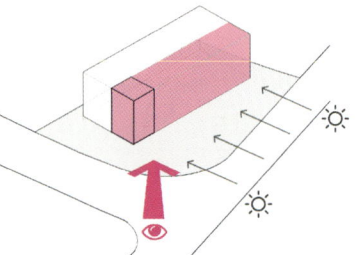

REFLECTING ARCH

Changing of flooring material, represents the reflection of shade and shadow from arch window on the floor.

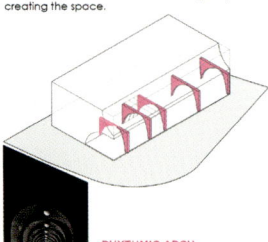

CROISSANT IMPERFECTION ARCH

The imperfection arch is formed for all spaces to create an art of imperfection

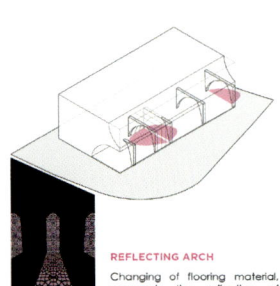

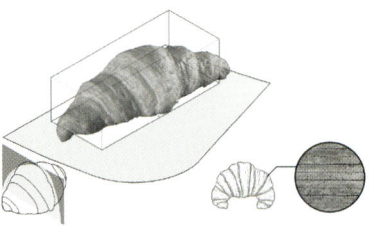

CROISSANT STRIPE MATERIAL

Using board formed concrete to create an imperfection texture, while displays as a croissant stripe.

DESIGN PROCESS

↑ Facade view   ↳ Partly view of exterior

프렌치 키치의 주된 디자인 콘셉트는 '프렌치(French)' 와 키치(Kitsch)' 의 합성어인 '프렌치 키치(French Kitsch)' 를 재해석하여 전개되었다. 프랑스 전문 파티시에라는 점 외에도 프렌치 불독에 대한 오너의 애정을 브랜드 아이덴티티로 삼아 카페에 유쾌하면서도 우아한 이미지를 담았다. 디자인은 건축적 요소와 재료를 통해 불완전함을 감상하는 예술의 한 형태인 '키치'의 콘셉트와 함께 이러한 이미지를 강화하는 것을 목표로 한다.

리드미컬한 아치가 발달한 프랑스 성당을 디자인에 주요 참고 자료로 삼았다. 대칭적인 아치 대신 다양한 스케일의 불완전한 아치가 사용되었다. 디자인은 완벽한 직사각형 덩어리에서 시작하여 1층에는 불완전한 아치를, 2층에는 반전된 불완전한 아치를 조각해냈다. 1층에서 이 아치들은 고풍스러우면서도 현대적인 모습으로 방문객을 감싸며 길을 따라 그림자를 만들고, 아치형 창에 빛이 통과하면 대성당의 유리처럼 바닥에 반사되는 효과를 만들어낸다. 또한 거대하고 불완전한 아치는 카운터에서 2층까지 연속적인 공간을 만들어 모든 기능을 갖춘 카운터를 강조하고, 두 층에서 모두 볼 수 있게 하였다. 2층에는 반전된 곡선에 보이드를 파내어 햇빛을 유도한다.

질감이 있는 콘크리트를 사용함으로써 벽면이 완전히 매끈하지는 않지만 그 자체로 아름다울 수 있는 재료의 진정성을 반영해 불완전함의 완벽함이라는 콘셉트를 강화했다. 또한 콘크리트를 주재료로 하여 녹색과 분홍색의 가구와 데코레이션, LED 조명이 돋보이도록 하여 브랜드의 컬러 아이덴티티를 강조함으로써 카페에 대한 강렬하고 기억에 남는 이미지를 연출하였다.

↑ Corner view

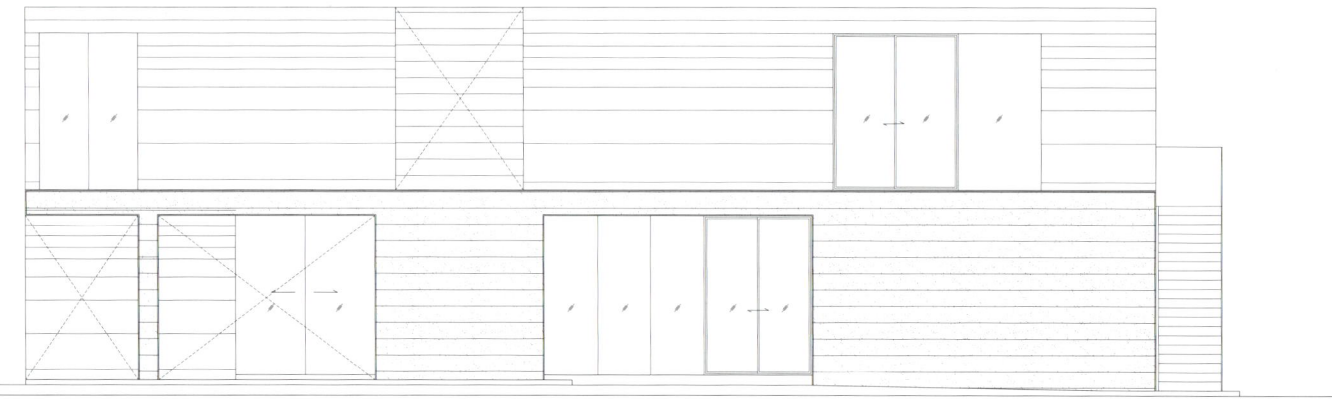

ELEVATION

' CROISSANT CURVES '
An imperfection arch curve from croissant's sectional cut

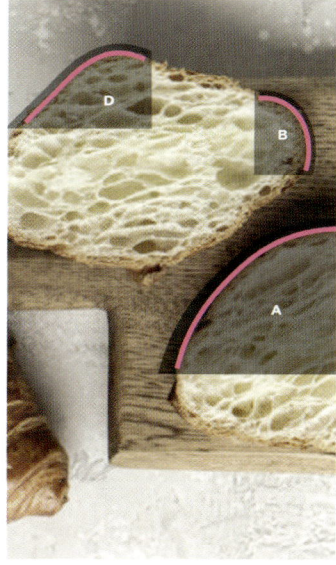

CEILING

A continuous linear curved ceiling over the overall interior space creates replicational curve from an exterior walls.

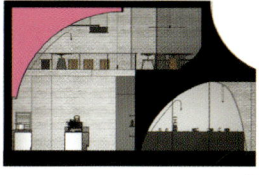

A

EXCAVATED FACADE

Eliminate an unused upper floor space by curving-cut, while creates a shading device and attractive facade.

B

INNER TUNNEL ARCH

The top arch is elevated down to create a slanted tunnel, which shapes a special space for sitting only.

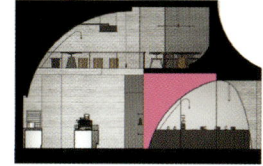

C

OUTER TUNNEL ARCH

Another half of the tunnel is asymmetrically formed to create an art of imperfection space inside.

D

CROISSANT CURVES

← Rhythmic arches   → Rhythmic arches

↑ Slow bar area ← Seating area → Seating area

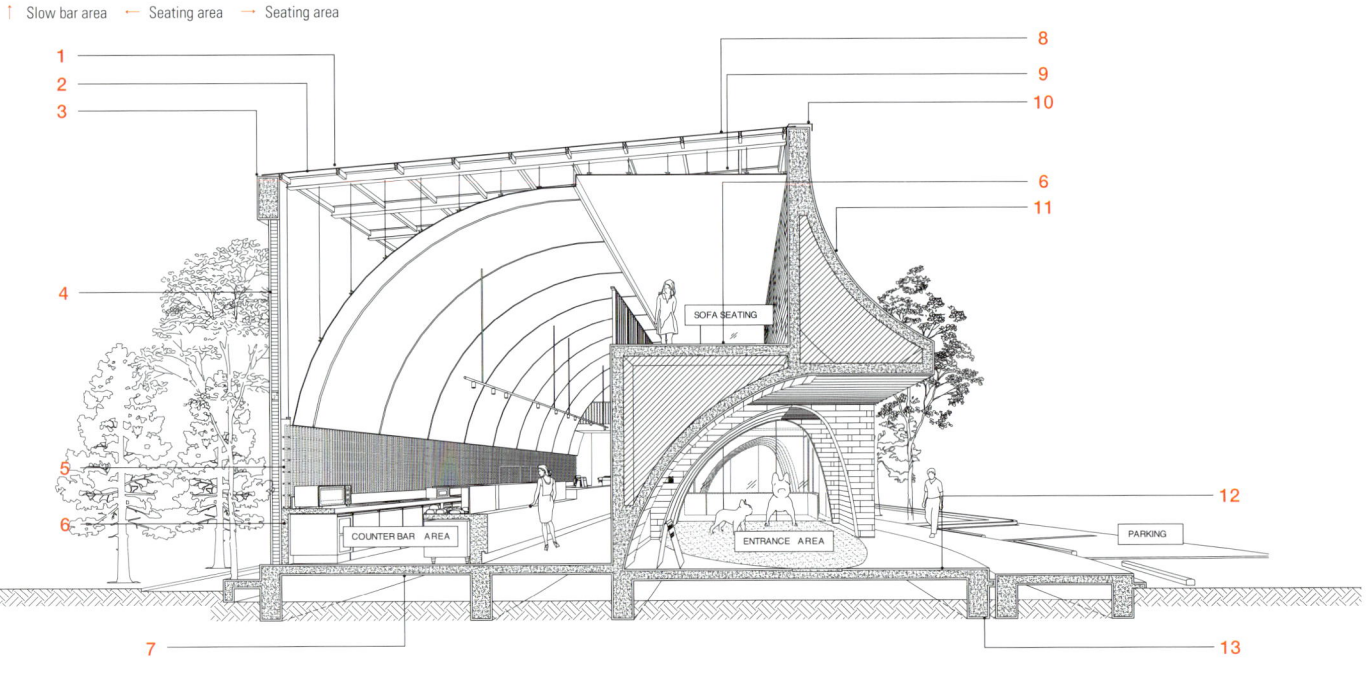

CROSS SECTION

← Counter bar   → Interior view

1. 50*50MM STEEL PIPES PURLIN
2. 200*150mm H-STEE RAFTER
3. 200*400MN REINFORCED CONCRETE ROOF BEAM
4. LIGHTWEIGHT CONCRETE WALL
5. 9MM BLACK STEEL GRATE
6. REINFORCED CONCRETE COUNTER WITH BLACK TERRAZZO FINISHING
7. REINFORCED CONCRETE SLAB WITH BLACK TERRAZZO FLOORING
8. METAL SHEET ROOF
9. T9MM GYPSUM BOARD CEILING WITH LIGHT-GREY PLASTERED AND PAINTED
10. ALUMINIUM FLASHING
11. REINFORCED CONCRETE WALL WITH WOODEN PATERN OF FORMWORK
12. REINFORCED CONCRETE SLAB WITH DARK-GREY EXPOSED AGGREGATE FINISHING
13. REINFORCED CONCRETE BEAM
14. TEMPERED GLASS WINDOW
15. LAMINATED-TEMPERED GLASS WINDOW
16. ACRYLIC SKYLIGHT ROOF
17. REINFORCED CONCRETE CEILING
18. 200*500mm REINFORCED CONCRETE BEAM

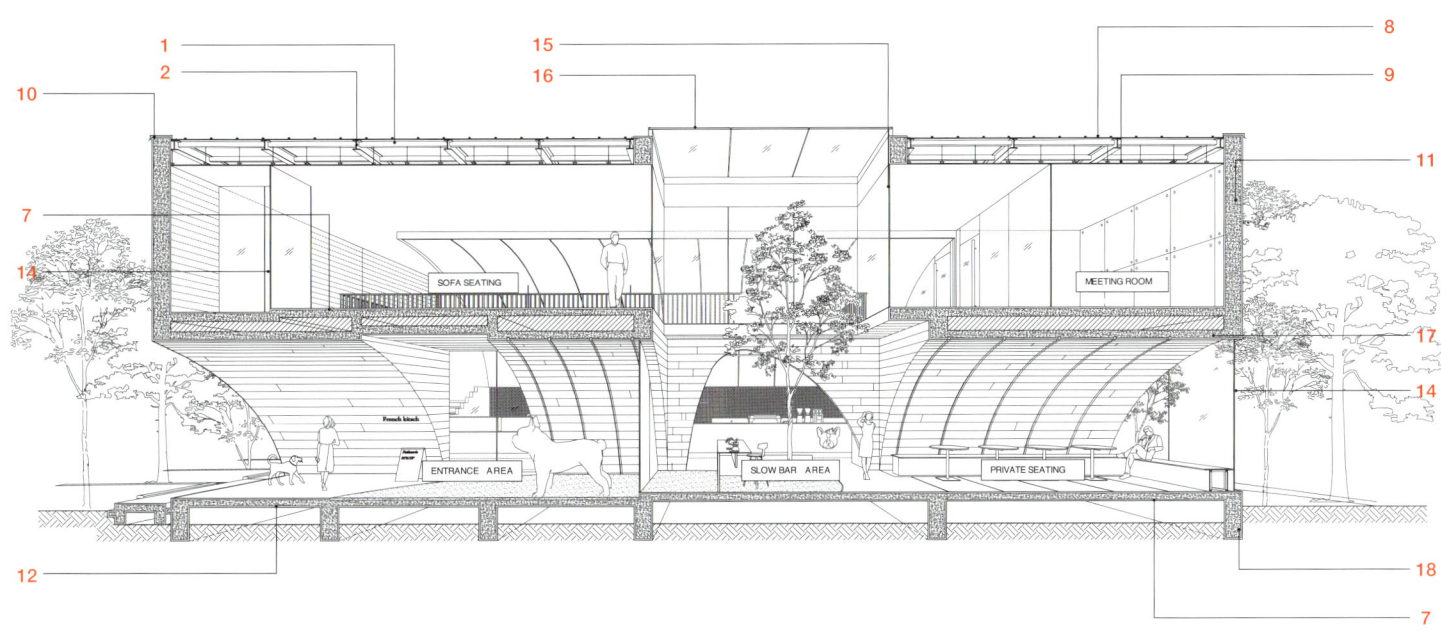

LONGITUDINAL SECTION

↑ Interior view ↓ Slow bar

↑ View of the 2nd floor

1	CPARKING LOT	4	COUNTER BAR	7	SLOW BAR	10	OUTDOOR STAIRCASE
2	ENTRANCE	5	SPEED BAR	8	SEATING AREA	11	MEETING ROOM
3	INDOOR STAIRCASE	6	STOCK ROOM	9	RESTROOM	12	SKYLIGHT

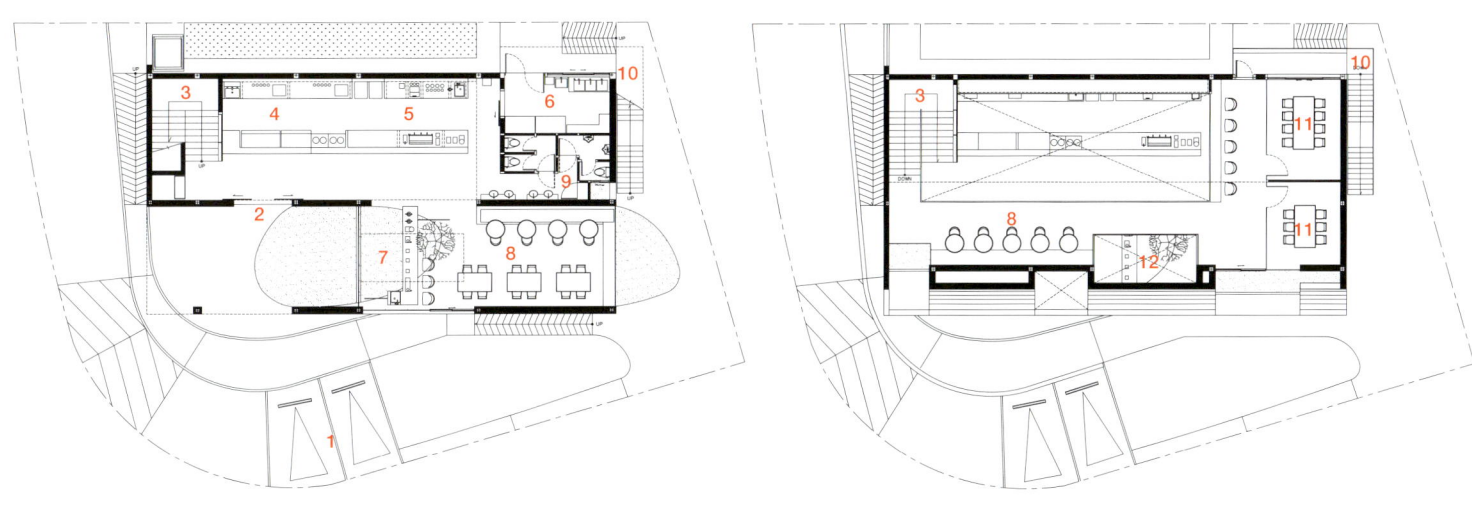

1ST FLOOR PLAN **2ND FLOOR PLAN**

포르투 바이앙 공공 도서관

PUBLIC LIBRARY IN BAIÃO, PORTO

ARCHITECT : TRAÇO ALTERNATIVO ARQUITECTOS ASSOCIADOS, lda. / NUNO CAMPOS

THE SITE OF THE NEW MUNICIPAL LIBRARY OF BAIAO FACES to the south with Av. April 25 and offers views of the landscape to the north. The intervention proposal was to rehabilitate and integrate the existing building and adding to it a new volume with three floors. This solution allowed to have wide glass windows facing North, taking advantage of the natural light, allowing ideal reading conditions.

The main entrance of the library, which gives access to the lobby, is made through the old primary school, as a way of honoring the past generations, bringing them to the present and projecting it into the future. The new generations would follow their footsteps and would also come here to acquire knowledge. A second entrance, through a covered side ramp, allows access to people with reduced mobility which creates a second atrium that functions as a small exhibition hall and, simultaneously, serves as a support space for the events to be held in the auditorium and in the training room. This alternative entrance adds versatility to the building, allowing it to have other uses when the library is closed, promoting and enabling the most diverse uses including cultural events.

The vertical access allows passage to the upper floor, to the adults' section, where people can watch the horizon and the silhouette of the mountains, as described in the book "The City and the Mountains" of Eca de Queiroz, written precisely about this untamed landscape. In the lower floor, we can find a warehouse, the internal service spaces and the children's section, that opens to the garden, which articulates with the inside to enable educational activities. The project tried to merge the two bodies in order to form a single entity, through formal and functional commitment and through the use of the same materials in the two volumes. Thus, the two bodies, although distant in time, seek the possible dialogue through volumetry, but also through materiality. Like the people who inhabit this region, who are extremely welcoming and transparent, the building is like them, with large windows and lots of light.

Location Baiao, Porto, Portugal **Use** Public Library **Site area** 1,055.0m² **Built area** 382.1m² **Gross Floor area** 1,065.7m² **Completion** 2023 **Project manager** Nuno Campos, Pedro Cardoso **Contractor** Fielnorte - Construcao E Engenharia Civil, Lda **Photographer** Alexander Bogorodskiy, Traço Alternativo arquitectos associados, lda.

OLD vs NEW

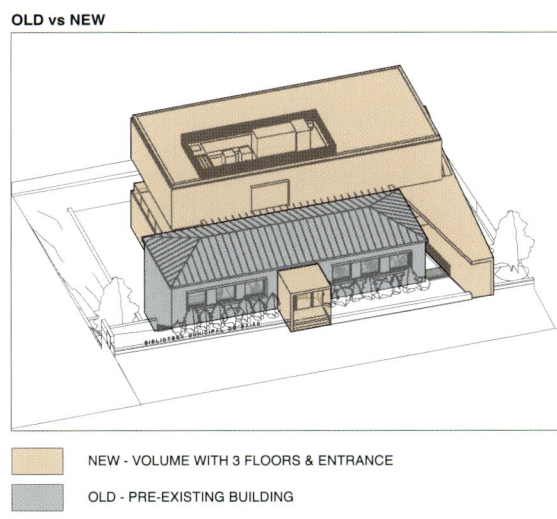

- NEW - VOLUME WITH 3 FLOORS & ENTRANCE
- OLD - PRE-EXISTING BUILDING

PROGRAM PER FLOOR

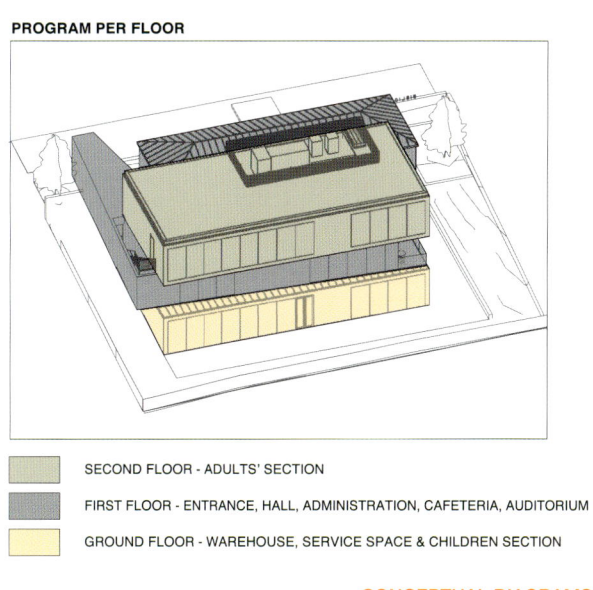

- SECOND FLOOR - ADULTS' SECTION
- FIRST FLOOR - ENTRANCE, HALL, ADMINISTRATION, CAFETERIA, AUDITORIUM
- GROUND FLOOR - WAREHOUSE, SERVICE SPACE & CHILDREN SECTION

CONCEPTUAL DIAGRAMS

← Bird's eyes view

↑ Corner view

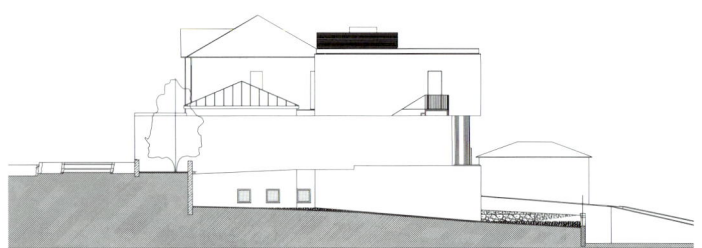

EAST ELEVATION

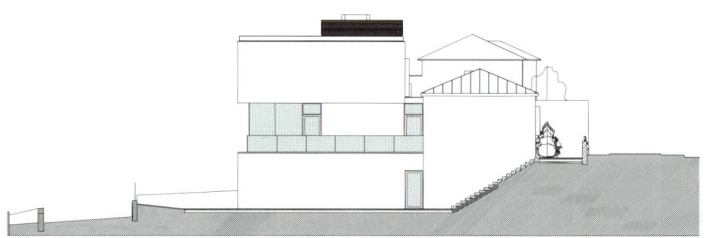

WEST ELEVATION

새로운 바이앙 공공 도서관(Public Library in Baião) 부지는 남쪽으로 에이프릴 25 에비뉴와 마주하고 있으며 북쪽의 풍경을 조망할 수 있다. 조정안은 기존 건물을 복구하고 통합하여 3층으로 된 새로운 건물을 추가하는 것이었다. 이 솔루션을 통해 북쪽을 향한 넓은 유리창을 설치하여 자연 채광을 활용하고 이상적인 독서 환경을 조성할 수 있었다.

로비로 통하는 도서관의 정문은 과거 세대를 기리고 현재를 이어주며 미래를 투영하는 의미에서 옛 초등학교를 통해 만들어졌다. 새로운 세대는 그들의 발자취를 따라 지식을 습득하기 위해 이곳에 오게 될 것이다. 두 번째 입구는 지붕이 있는 측면 경사로를 통해 거동이 불편한 사람들도 출입할 수 있으며, 작은 전시장 역할을 하는 두 번째 아트리움을 만들어 강당과 교육실에서 열리는 행사를 위한 지원 공간으로 활용할 수 있다. 이 대체 출입구는 건물에 다목적성을 더하여 도서관이 문을 닫았을 때 다른 용도로 사용할 수 있도록 하여 문화 행사를 비롯한 다양한 용도로 활용할 수 있도록 홍보하고 활성화한다.

수직 출입구는 이 길들여지지 않은 풍경에 대해 정확하게 묘사한 에카 데 케이로즈(Eca de Queiroz)의 저서 '도시와 산(The City and the Mountains)'에 묘사된 것처럼 사람들이 지평선과 산의 실루엣을 볼 수 있는 위층, 성인용 섹션으로 이동할 수 있게 해준다. 아래층에는 창고, 내부 서비스 공간 및 어린이 섹션이 있으며 정원으로 연결되는 창고가 있으며 내부와 연결되어 교육 활동이 가능하다. 이 프로젝트는 형식적, 기능적 헌신과 두 권에 동일한 재료를 사용하여 두 기관을 통합하여 단일 개체를 형성하려고 시도했다. 따라서 두 기관은 시간적으로 멀리 떨어져 있지만 부피 측정뿐만 아니라 물질성을 통해 가능한 대화를 추구한다. 매우 따뜻하고 투명한 이 지역에 사는 사람들처럼 건물은 큰 창문과 많은 빛으로 그들을 닮았다.

↑ Exterior view

SOUTH ELEVATION

NORTH ELEVATION

↑ A three-dimensional signage   ↲ Exterior view

↑ Night view

SECTION

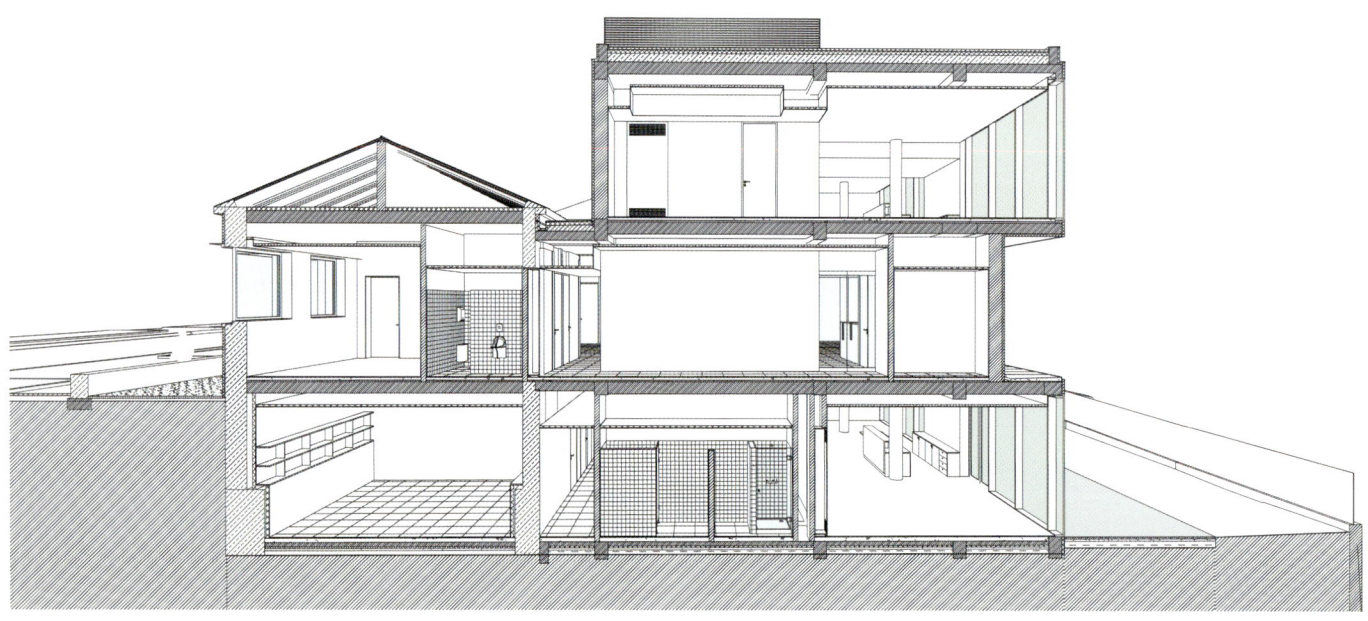

↑ Reading area   → Staircase

1   60mm WASHED GRAVEL
2   GEOTEXTILE
3   XPS 80mm THERMAL INSULATION
4   WATERPROOFING MEMBRANE
5   LAYER WITH INTEGRATED SLOAP >2%
6   SLAB
7   ZINC CASING
8   POLYETHYLENE MEMBRANE
9   CONCRETE UPSTAND
10 XPS 40mm THERMAL INSULATION
11 ZINC
12 GYPSUM PLASTERBOARD
13 STONE WOOL
14 ACRYLIC PLASTER
15 EPS 60mm THERMAL INSULATION
16 THERMO-LACQUERED ALUMINUM WINDOW FRAME
17 WOODEN FLOOR
18 100mm FILLING AND REGULARIZATION LAYER WITH ACOUSTIC MEMBRANE
19 SHORT CONCRETE WALL
20 10X10mm DRIP GROOVE
21 THERMO-LACQUERED ALUMINUM GRID PAINTED WHITE
22 CONCRETE BEAM
23 300mm THERMAL BLOCKS
24 TINNED FINISH PLASTER
25 ZINC-PLATED ROOF - SEALED JOINT
26 110mm HOLLOW BRICK MASONRY
27 15mm OSB PANEL
28 PRETENSIONED JOISTS
29 RESTORED GRANITE CORNICE
30 EPS 100mm THERMAL INSULATION
31 ZINC GUTTER
32 GLAZED PORCELAIN
33 SMOKE EXTRACTOR
34 XPS 60mm THERMAL INSULATION
35 LED STRIP LIGHT FACING DOWN
36 WOOD STRUCTURE TO ELEVATE AND FIXATE THE LED STRIP LIGHT
37 CONCRETE WALL
38 CLOSET

SECTION

SECTION DETAIL

← Interior view → Staircase ↵ Elevator hall

1 READING AREA FOR CHILDREN
2 PHOTOCOPIER RECEPTION
3 RESTROOM
4 STORAGE
5 COMPACT SHELVES
6 OFFICE
7 MAIN ENTRANCE
8 RECEPTION
9 CAFETERIA
10 ATRIUM
11 AUDITORIUM
12 TRAINING ROOM
13 READING AREA FOR ADULT
14 PERIODICAL ZONE

↑ Reading area

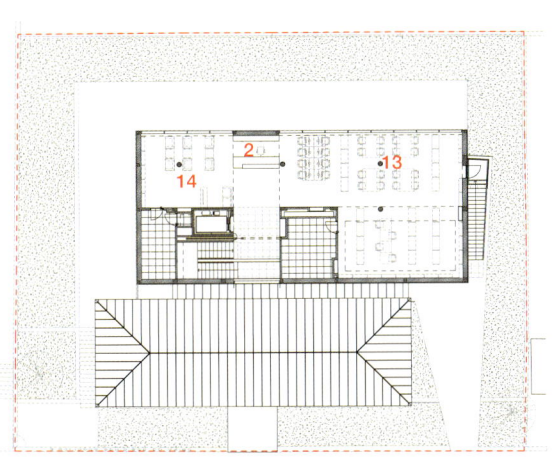

2ND FLOOR PLAN

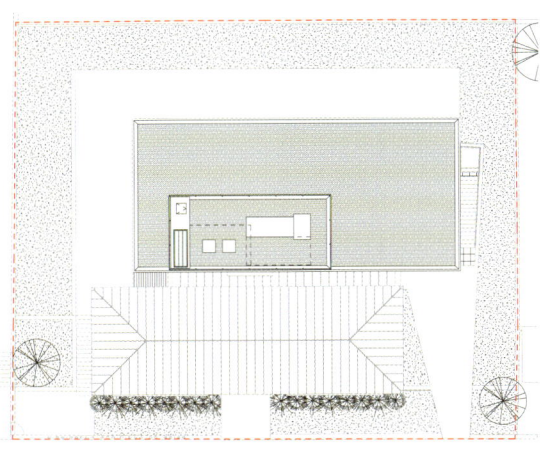

ROOF FLOOR PLAN

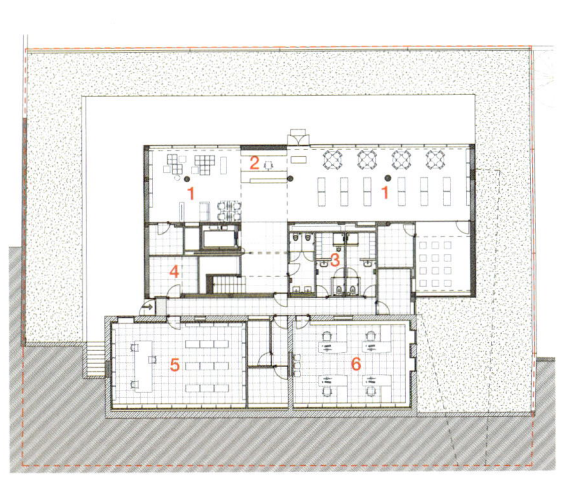

GROUND FLOOR PLAN

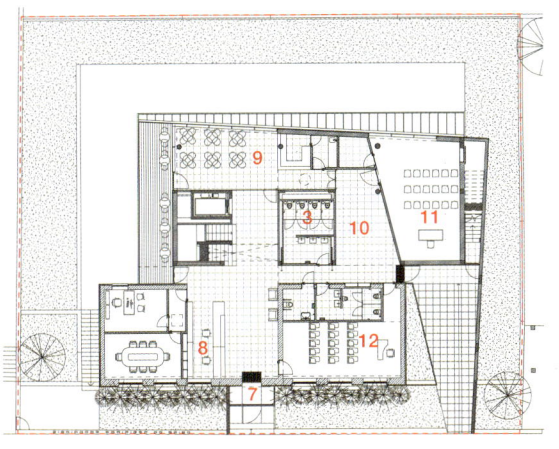

1ST FLOOR PLAN

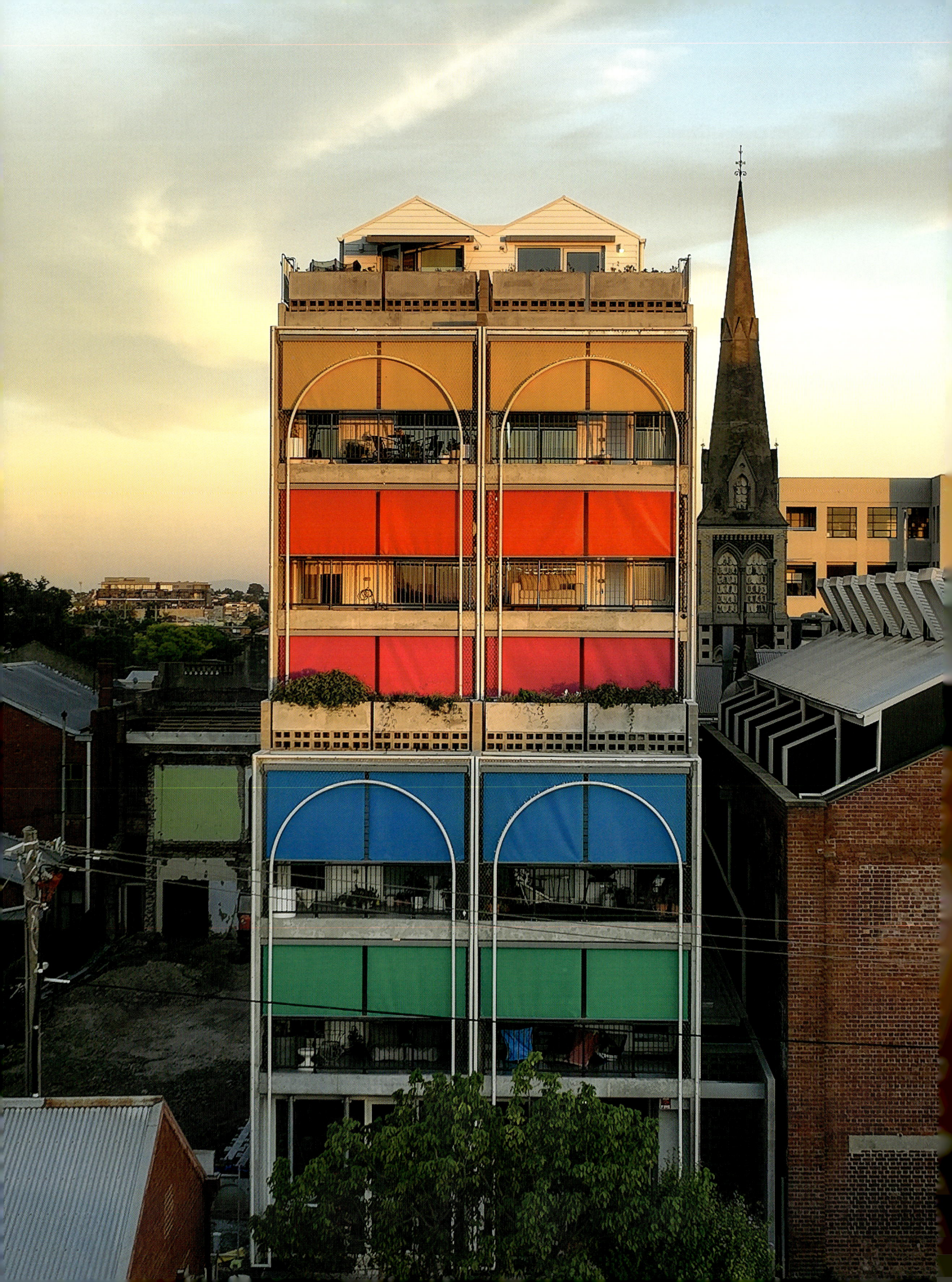

테라스 하우스
TERRACE HOUSE

ARCHITECT : AUSTIN MAYNARD ARCHITECTS / ANDREW MAYNARD, MARK AUSTIN, MARK STRANAN

TERRACE HOUSE IS AN ETHICAL, BEAUTIFULLY-DESIGNED, highly sustainable and 100% fossil fuel free building in Melbourne, Australia. Terrace House takes a revolutionary approach to housing and delivers community-focused, environmentally, socially and financially sustainable homes that are robust and resilient in the face of the growing climate crisis.

Located on a busy high street, close to all amenities, the building comprises of 20 (two and three bedroom) residences, with 55 bike parks and three commercial spaces. Intended as owner-occupier, Terrace House is the re-imagining of a former inner-city suburban life, where rows of workers cottages generated and nurtured close community. Shared childminding, communal gardens, neighbourly lending and borrowing - these ideals are the basis of Terrace House. These are not apartments but terrace houses, stacked six storeys high.

The average Australian home measures 233m² and is, typically poorly designed, high maintenance and inefficient in terms of space and energy. By contrast, small inner-city apartment buildings seek to maximise returns, using saddlebacks, compromising bedrooms and facing homes inwards, towards each other. Austin Maynard Architects believe homes should have an aspect out from the site, into the surroundings and open sky. Responding to this unique site (a long block measuring 10mx57m) we took the opportunity to emulate traditional terrace house plans. Homes with big external outlooks, a front verandah, study and shared 'backyard' on the roof.

Terrace House seeks to be a positive example of good urbanism, focusing on deliberative design over profit and working together with residents to inform the outcome and help them author their collective future. With the support, belief and funding of past clients and allies, and at huge personal risk to the directors and team members, the practice adopted the role of both architect and developer - to create homes focused on long-term liveability, where ongoing costs were a priority, not an afterthought.

In Australia, multi residential buildings are require to achieve 6 stars within the building code certification method. Terrace House is 8.1. Internationally equivalent to a 6-star Green Star rating (considered world leading) and 'Platinum' LEED and BREEAM 'Outstanding' classification.

The form of the street facades of Terrace House is a direct response to this locality's rich and diverse built heritage. The area has many examples of post war Mediterranean-Australian architecture, industrial buildings and grand Victorian shop fronts. Terrace House borrows from this context in a respectful and playful way. A modern interpretation of the context.

Location Melbourne, Australia **Use** Housing **Site area** 597m² **Gross floor area** 3,225m² **Project manager** Armitage Jones **Developer** Austin Maynard Architects **Builder** Kapitol Group **Engineers** Structural - Adams Engineering, Engineer - Services - BCA Engineers, Engineer - Fire - Omnii, Engineer - Acoustic - WSP **ESD** Irwin Consult **Traffic Consultant** Ratio **Planning consultant** Hansen **Access Consultant** Architecture and Access **Landscape Architects** Openwork **Building Surveyor** Steve Watson & Partners **Stairwell graphics** Hermann Studios **Photographer** Derek Swalwell

← Exterior view at sunset   → Exterior night view

← West view & colorful blind   → Extrior view & entrance

EAST ELEVATION

EAST ELEVATION (WITH BLIND DOWN)

↑ East side exterior view

WEST ELEVATION

WEST ELEVATION (WITH BLIND DOWN)

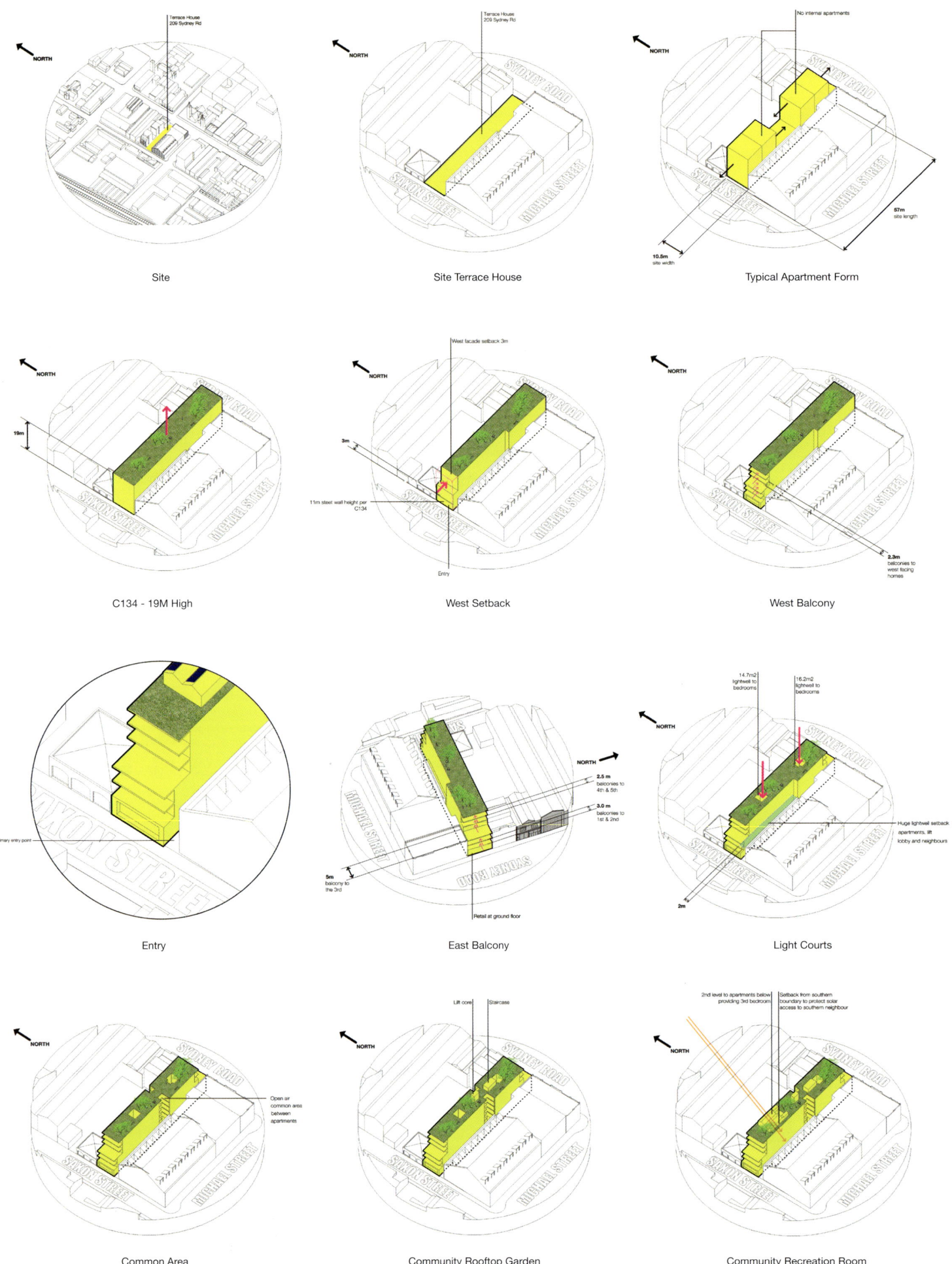

DIAGRAM

→ West, street view

테라스 하우스는 호주 멜버른에 위치한 윤리적이고 아름답게 디자인된, 지속 가능하며 100% 화석 연료를 사용하지 않는 건축물이다. 테라스 하우스는 주택에 대한 혁신적인 접근법을 취하며, 환경적, 사회적, 경제적으로 지속 가능한 주택을 제공한다. 이는 강력하고 회복력이 있어 점점 커지는 기후 위기에도 견딜 수 있도록 공동체 중심의 가치를 담고 있다.

건물은 번화가에 위치해 모든 편의시설이 가깝고 20개의 (두세 개의 침실을 갖춘) 주거 공간과 55대의 자전거 주차 공간 및 3개의 상업공간이 있다. 소유주 거주가 목적이며 테라스 하우스는 내부 도시 교외의 삶을 새롭게 상상하며, 노동자의 오두막이 줄지어 있던 과거의 커뮤니티의 밀접함을 생성하고 육성하였다. 공동 육아, 공동 정원, 이웃 간의 물품 대여 및 빌려주기 이러한 이상은 테라스 하우스의 기초이다. 이곳의 주택들은 단순한 아파트가 아니라, 여섯 층 높이로 쌓인 테라스 하우스이다.

평균적인 호주 주택은 233㎡의 면적을 가지고 있으며, 흔히 설계가 미흡하고, 유지 관리가 많이 들며, 공간과 에너지 측면에서 비효율적이다. 반면, 소규모 도심 아파트 건물은 최대의 수익을 내기 위해 새들백을 사용하고 침실의 크기를 타협하여 주택들을 서로 안쪽을 향하도록 배치하였다. 건축가는 주택들이 그 장소를 벗어나 주변 환경과 개방된 하늘로 조망을 추구하였다. 길고 좁은 블록(10m x 57m)이라는 독특한 장소에 대응하여, 전통적인 테라스 하우스의 평면도를 모방할 기회를 가졌다. 큰 외부 조망을 가진 집들, 앞 베란다, 작업실, 그리고 지붕 위의 공유와 '뒷마당'을 갖춘 주택이다.

테라스 하우스는 수익보다는 신중한 설계에 초점을 맞추고, 거주민들과 협력하여 결과를 알리고 공동의 미래를 함께 만드는 좋은 도시주의의 긍정적인 본보기가 되고자 했다. 과거의 고객과 동맹자들의 지지, 믿음, 그리고 자금 지원을 받아, 그리고 이사진과 팀원들의 큰 개인적 위험이 큰 이 관행은 건축가와 개발자의 역할을 채택하여 장기적인 생활성에 중점을 둔 집들을 만들었다. 지속적인 비용이 사후 처리가 아닌 우선순위라 믿었다.

호주에서는 다세대 주택 건물이 건축 코드 인증 방식에 따라 6성급을 달성해야 한다. 테라스 하우스는 그 기준을 뛰어넘어 8.1의 평가를 받았다. 이는 국제적으로 6성급 그린 스타 등급(세계적인 수준으로 간주)과 동등하며, '플래티넘' 리드와 브리암 '아웃스탠딩' 분류에 해당한다.

테라스 하우스의 거리 정면 형태는 이 지역의 풍부하고 다양한 건축 유산에 대한 직접적인 반응이다. 이 지역에는 전후 지중해-호주 건축, 산업 건물 및 웅장한 빅토리아 시대의 상점 정면 등 많은 예시가 있다. 테라스 하우스는 이러한 맥락을 존중하고 유쾌한 방식으로 차용한다. 이는 맥락에 대한 현대적 해석이다.

← East entrance   → Back side garden & entrance

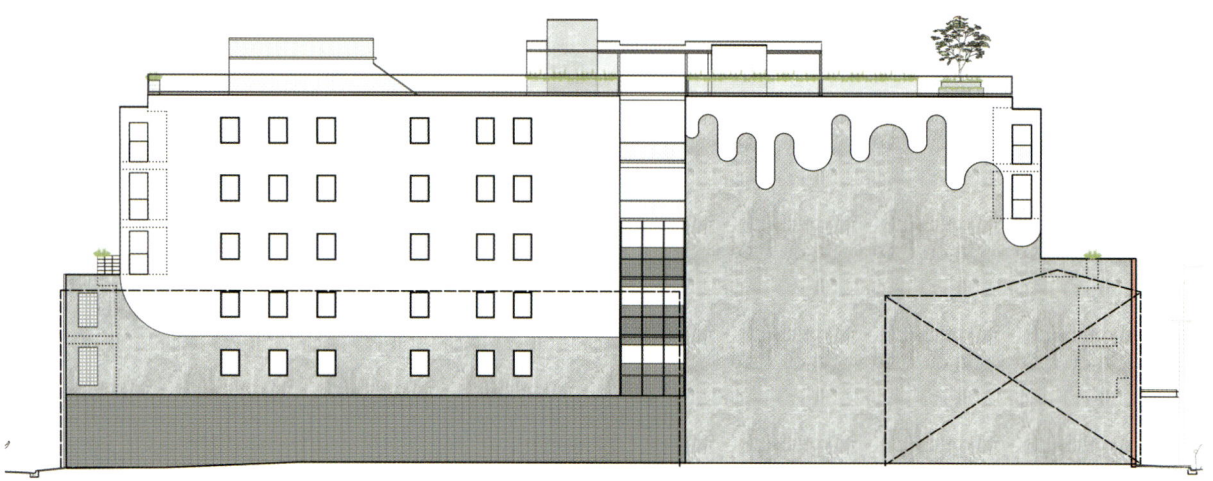

SOUTH ELEVATION

← Look up view → Common area

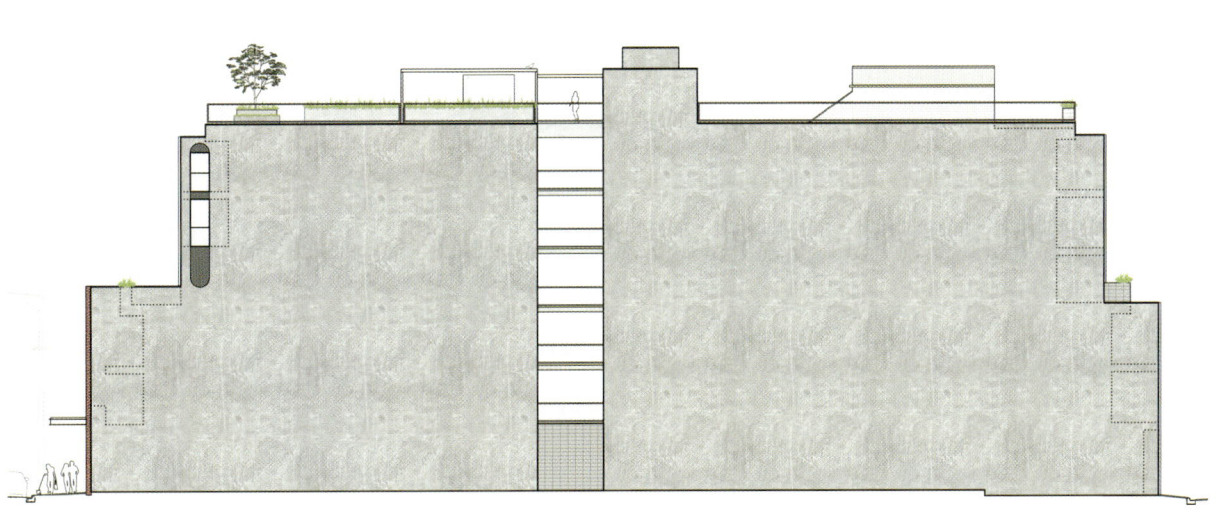

NORTH ELEVATION

↑ Veranda ↙ Balcony ↓ Balcony ↘ Balcony

↖ Veranda ↗ Veranda & rooftop balcony ↙ Living room & balcony ↓ Balcony ↘ Common area

↑ Living room   ↵ Kitchen   ↓ Bedroom   ↳ Living room & dining room

← Living room   → Living room & balcony

1 SHARED COMMON AREA	5 TOILET	9 LIGHTWELL	13 STORAGE
2 VERANDA	6 BEDROOM	10 GARDEN BED	14 ELEVATOR
3 ENTRY / STUDY	7 LIVING ROOM / DINING ROOM	11 BICYCLE PARKING	15 LAUNDRY
4 HALLWAY	8 BALCONY	12 BACK OF HOUSE	

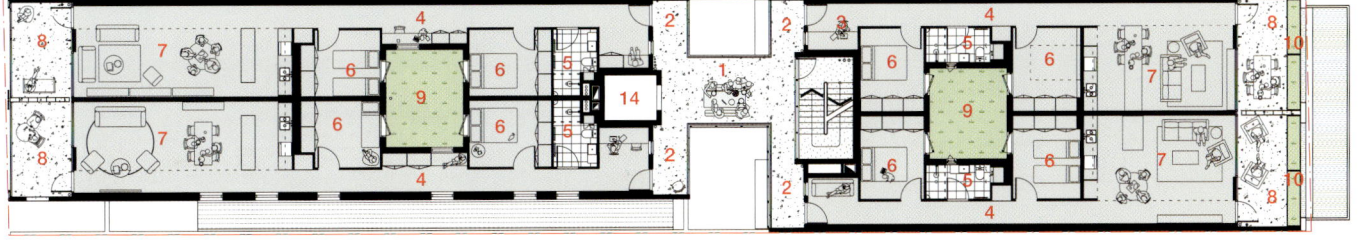

1ST FLOOR PLAN

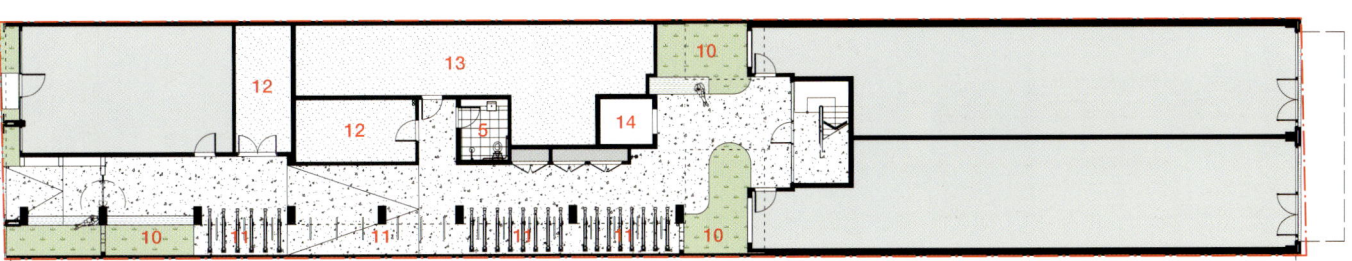

GROUND FLOOR PLAN

↖ Staircase   ↑ Staircase wall graphic   ↗ Hallway

1 SHARED COMMON AREA	5 TOILET	9 LIGHTWELL	13 STORAGE
2 VERANDA	6 BEDROOM	10 GARDEN BED	14 ELEVATOR
3 ENTRY / STUDY	7 LIVING ROOM / DINING ROOM	11 BICYCLE PARKING	15 LAUNDRY
4 HALLWAY	8 BALCONY	12 BACK OF HOUSE	

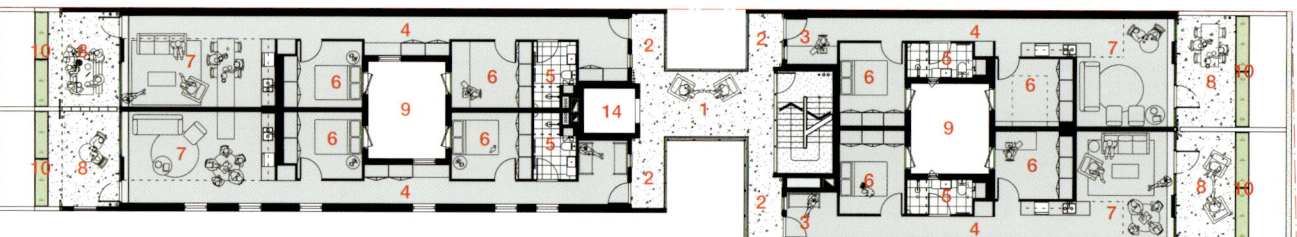

3RD FLOOR PLAN

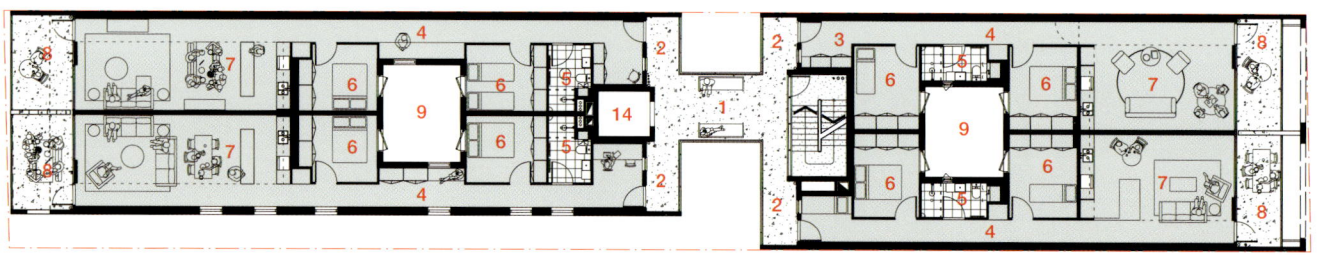

2ND FLOOR PLAN

← Rooftop balcony   ⌐⌐ Rooftop common area   ⌐ Rooftop garden

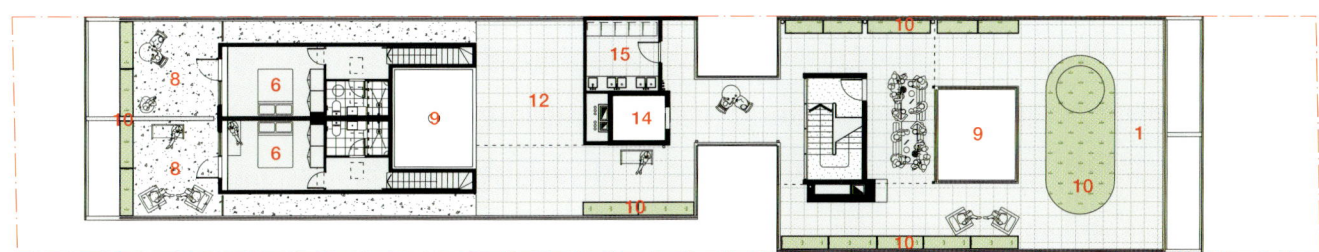

ROOF TOP FLOOR PLAN

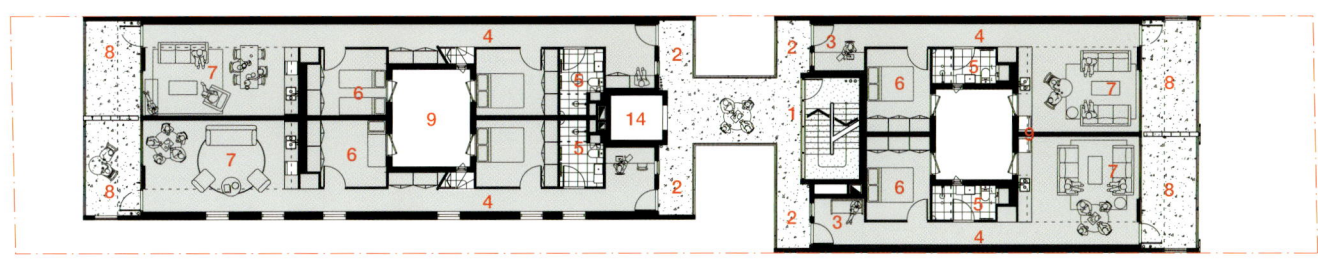

5TH FLOOR PLAN

시티 프레시
CITY FRESH

ARCHITECT : SPACY COMPANY LIMTED / PRAMOTH KITKANASIRI

CITY FRESH WAS A REGULAR ROW HOUSE that has been transformed into a space that welcomes fruit lovers. The venue is somewhat a destination in which groups of people with common interests can gather, exchange, and share. Space on the first level has been prepared for brief and small meetings. There are temporary seatings with wide variety of curated fruits and fresh home grown salads to choose from. House of fruits incorporates different low-key transparent materials within structure to add natural light into the space and build layers that bring about comfort visual textures. Clear glass panels in the front interchangeably establishes connection and cohesivity among the architecture and its adjacencies. Open space on the first level concurrently help link and draw circulation from first to the upper levels.

Secondary skin on the structure differentiates public from private space. The element incorporates transparency with randomized void to make visual frames, creating different experiences and significant point of views once sitting in. Natural light passing through the facade softens modern essence by making layers of visibility and shades to greenery. Interior wall finishes continue textural visual element and language from the exterior. Neutral materials allow colors of fresh fruits to shine and standout. Space on the second floor features types of tables and seatings to cater different activities. Common space connected within the area encourages interaction among people.

Cooking studio with movable kitchenette on the third level allows for people to meet new faces and make friends. Cooking classes prepared by professional and well-known chefs are open for ones who wished to create healthy cooking lifestyle by incorporating vegetables or fruits to the dishes. Students are also able to pick fresh materials from the store on the ground level to create particular menus. The flexible cooking area is connected to private dining space which can be joined and converted to accommodate group activities. Lucky tree located in the middle does not only make center of attention to the area and structure as a whole, but also create shade and comfortability for employees while working at the top floor.

Location Bangkok, Thailand **Use** Retail, Cafe & Restaurant **Site area** 214m² **Built area** 125m² **Gross area** 602m² **Completion** 2022 **Design team** Pramoth Kitkanasiri, Patitta Khayan **Contractor** Yoohui Interior Co., Ltd. **Photographer** Rungkit Charoenwat

ELEVATION

← Exterior night view → Exterior view

↑ Exterior & street view

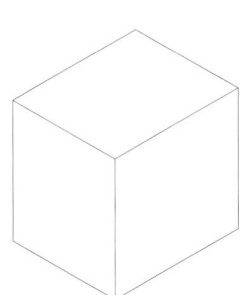
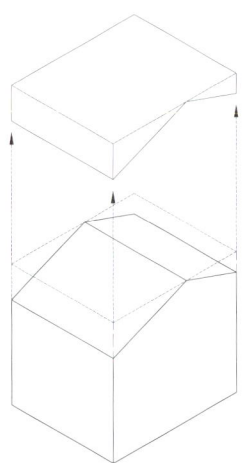
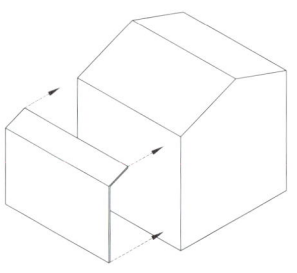
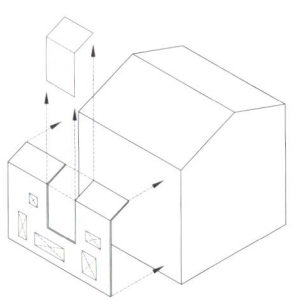
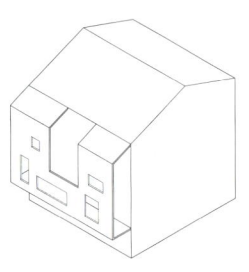
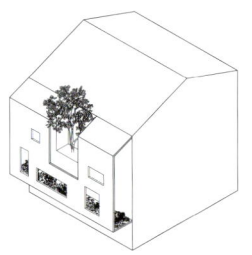

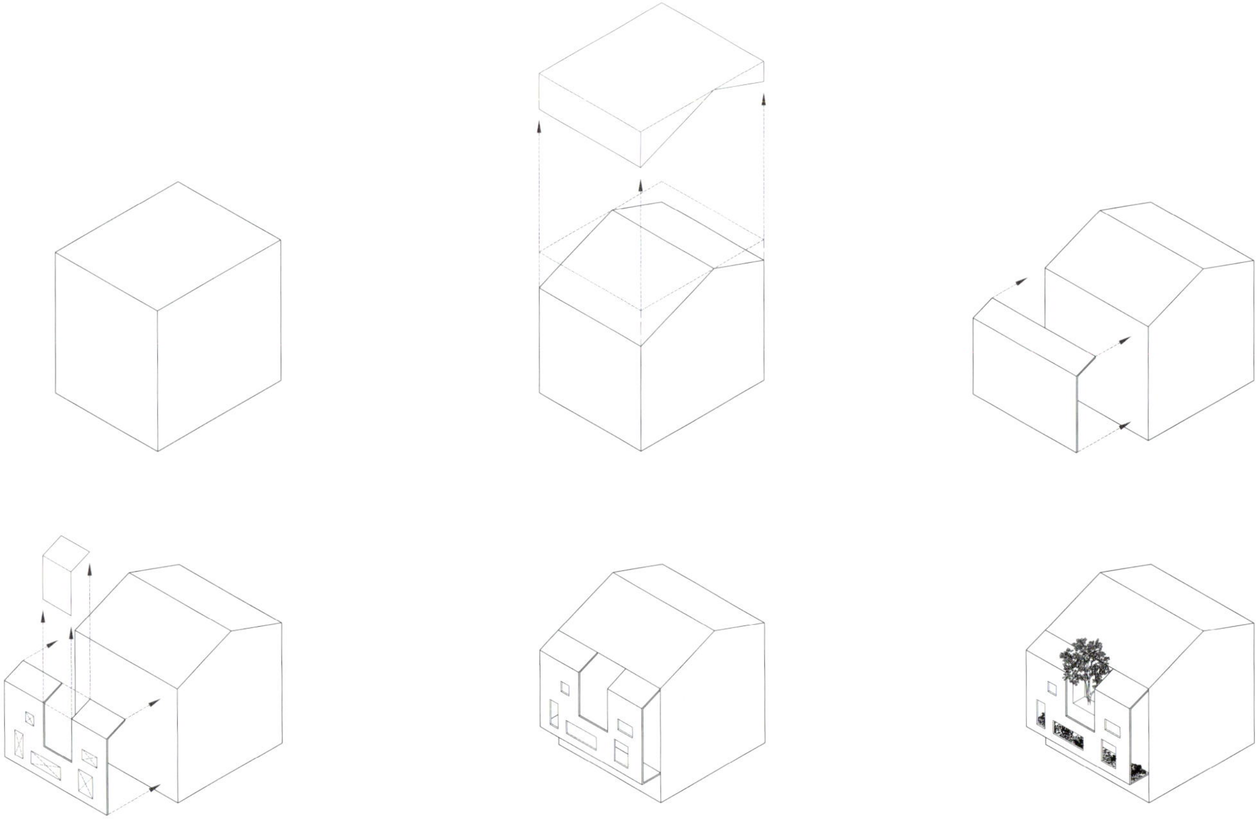

DIAGRAM

↑ Top view   ↳ Exterior night view

시티 프레시는 일반 연립주택을 과일 애호가들이 환영하는 공간으로 탈바꿈하였다. 이 장소는 공통된 관심사를 가진 사람들이 모여 교류하고 공유할 수 있는 공간이다. 1층 공간은 간단하고 소규모의 모임을 위한 공간이다. 다양한 종류의 엄선된 과일과 신선한 가정식 샐러드를 선택할 수 있는 임시 좌석이 마련되어 있다. 하우스 오브 프루츠는 구조 내에 다양한 로우키(low-key) 투명 재료를 통합하여 자연광을 안으로 유입시키고, 시각적인 질감을 제공하는 레이어를 구축하였다. 전면의 투명한 유리 패널은 건축물과 그 주변과의 연결성 및 일체감을 상호교환적으로 설정했다. 1층의 개방된 공간은 상위층까지 동선을 연결하고 유도하는 역할을 동시에 수행한다. 구조물에 추가된 2차 외벽은 공공 공간과 사적 공간을 구분했다. 이 요소는 무작위의 공간을 통해 투명성을 가미하여 시각적 프레임을 만들고, 내부에 앉았을 때 다양한 경험과 중요한 관점을 창조한다. 파사드를 통해 들어오는 자연광은 가시성과 녹지에 대한 음영 층을 만들어 현대적 본질을 부드럽게 만든다. 내부 벽 마감은 외부에서의 질감 있는 시각적 요소와 언어를 이어간다. 뉴트럴 소재를 사용하면 신선한 과일의 색상이 빛나고 돋보이게 할 수 있다. 2층의 공간은 다양한 활동을 수용할 수 있도록 여러 형태의 테이블과 좌석이 있다. 공용 공간은 지역 내에서 상호 작용을 장려하는 연결 고리를 형성한다. 3층에 위치한 요리 스튜디오는 이동식 간이 주방이 있고 사람들은 새로운 얼굴을 만나고 친구를 사귈 수 있다. 전문가이자 유명한 셰프들이 준비한 요리 수업은 채소나 과일을 요리에 곁들여 건강한 요리 생활을 만들고자 하는 이들을 위해 열려 있다.

학생들은 또한 1층에 있는 상점에서 신선한 재료를 직접 골라 특정 메뉴를 만들 수도 있다. 유연한 요리 공간은 개인 식사 공간과 연결되어 있으며, 이는 그룹 활동을 수용할 수 있도록 결합하거나 변형할 수 있다. 가운데 위치한 행운 나무는 공간과 구조 전체의 이목을 집중시킬 뿐만 아니라 최상층에서 일하는 직원들에게 그늘과 편안함을 제공한다.

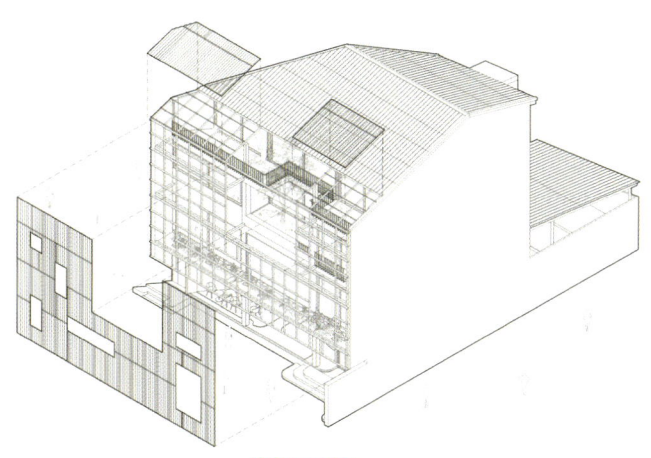

ISOMETRIC

↑ The restaurant entrance night view   ↓ 1st floor restaurant & parking

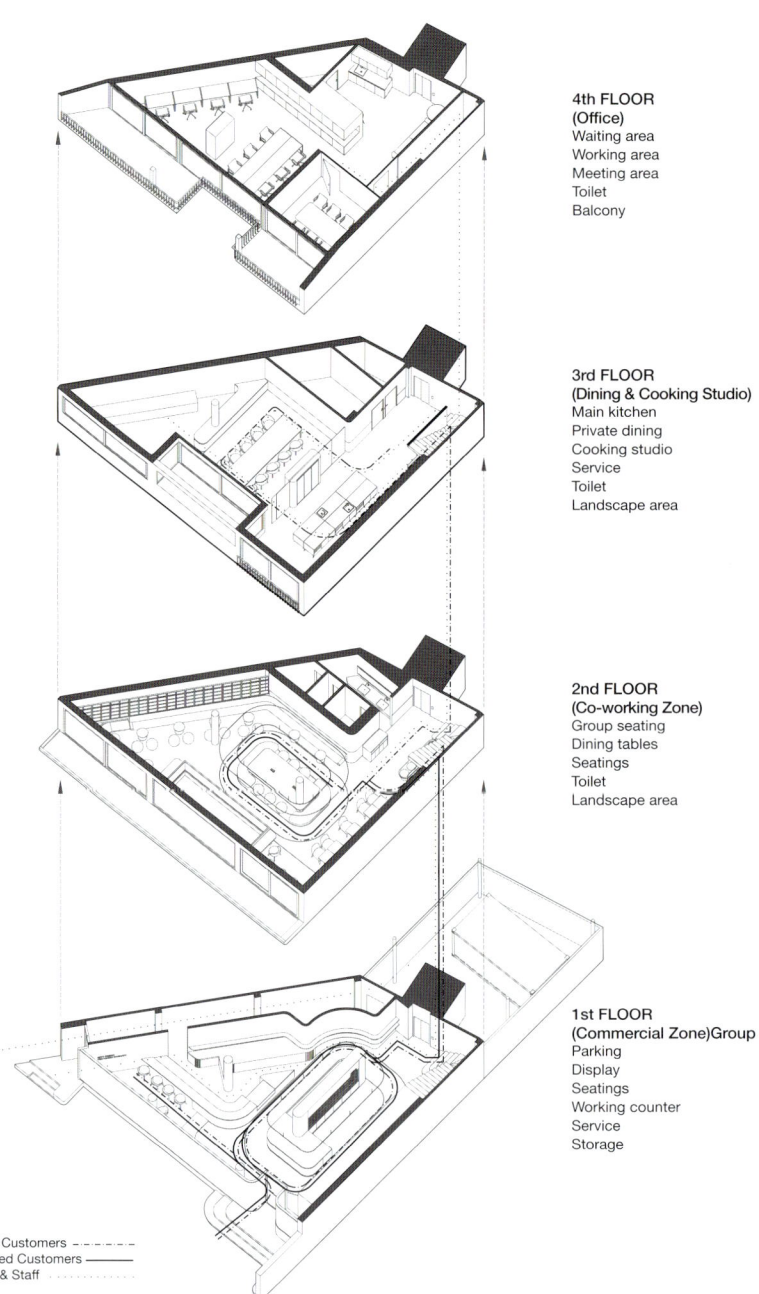

4th FLOOR
(Office)
Waiting area
Working area
Meeting area
Toilet
Balcony

3rd FLOOR
(Dining & Cooking Studio)
Main kitchen
Private dining
Cooking studio
Service
Toilet
Landscape area

2nd FLOOR
(Co-working Zone)
Group seating
Dining tables
Seatings
Toilet
Landscape area

1st FLOOR
(Commercial Zone)Group
Parking
Display
Seatings
Working counter
Service
Storage

Regular Customers -------
Appointed Customers ———
Service & Staff ....

CIRCULATION DIAGRAM

↑ Facade, glass panels in the front

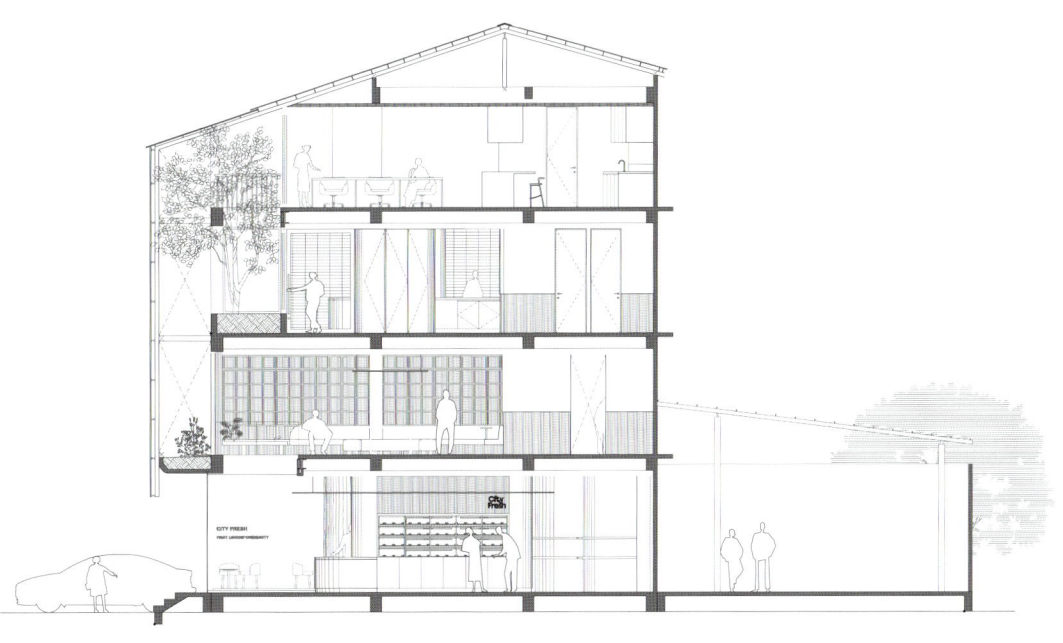

SECTION

← Facade detail & sign   → Facade detail, glass panels in the front   ↙ Facade detail & 2nd floor window   ↘ Facade detail

1 INSULATED METAL SHEET ROOFING (WHITE)
2 STEEL ROOF FRAME WITH ENAMEL PAINT
3 STEEL ROOF FRAME WITH ENAMEL PAINT
4 PLASTER CEILING WITH ROCKWOOL INSULATION
5 TRANSPARENT FIBER GLASS ROOF
6 INTERIOR WASHABLE PAINT (WHITE)
7 WHITE ALUMINUM WINDOW FRAME WITH 12mm
8 TEMPERED LAMINATED GLASS
9 CUSTOMIZED STEEL HANDRAIL
10 EXISTING REINFORCED CONCRETE STRUCTURE
11 STEEL FACADE FRAME WITH ENAMEL PAINT (WHITE)
12 WHITE ALUMINUM WINDOW FRAME WITH 12mm
13 TEMPERED LAMINATED GLASS
14 INTERIOR WHITE CEREMIC TILES
15 STEEL FACADE FRAME WITH ENAMEL PAINT (WHITE)
16 INTERIOR BUILT-IN FURNITURE
17 STEEL FACADE FRAME WITH ENAMEL PAINT (WHITE)
18 LANDSCAPE GRAVEL AND PLANTING SOIL
19 WATER PROOFING MEMBRANE
20 EXISTING REINFORCED CONCRETE STRUCTURE
21 PLASTER CEILING
22 CUSTOMIZED INTERIOR PENDANT (WHITE)
22 INTERIOR FEATURED WALL WITH WHITE SPRAY PAINT
22 TEMPERED LAMINATED GLASS 12mm. THICK
23 FRAMELESS TEMPERED LAMINATED GLASS (12mm) HANDRAIL
24 WATER PROOFING MEMBRANE
25 GLASS SUPPORT (U CHANNEL)
26 INTERIOR CURTAIN
27 CUSTOMIZED INTERIOR PENDANT (WHITE)
28 INTERIOR FEATURED WALL WITH WHITE SPRAY PAINT
29 ARTIFICIAL STONE INTERIOR FINISHING
30 TEMPERED LAMINATED GLASS 15mm. THICK
31 POLISHED CONCRETE WITH ANTI-SLIP COATING
32 LED STRIP LIGHT

↑ 1st floor restaurant counter & seating area

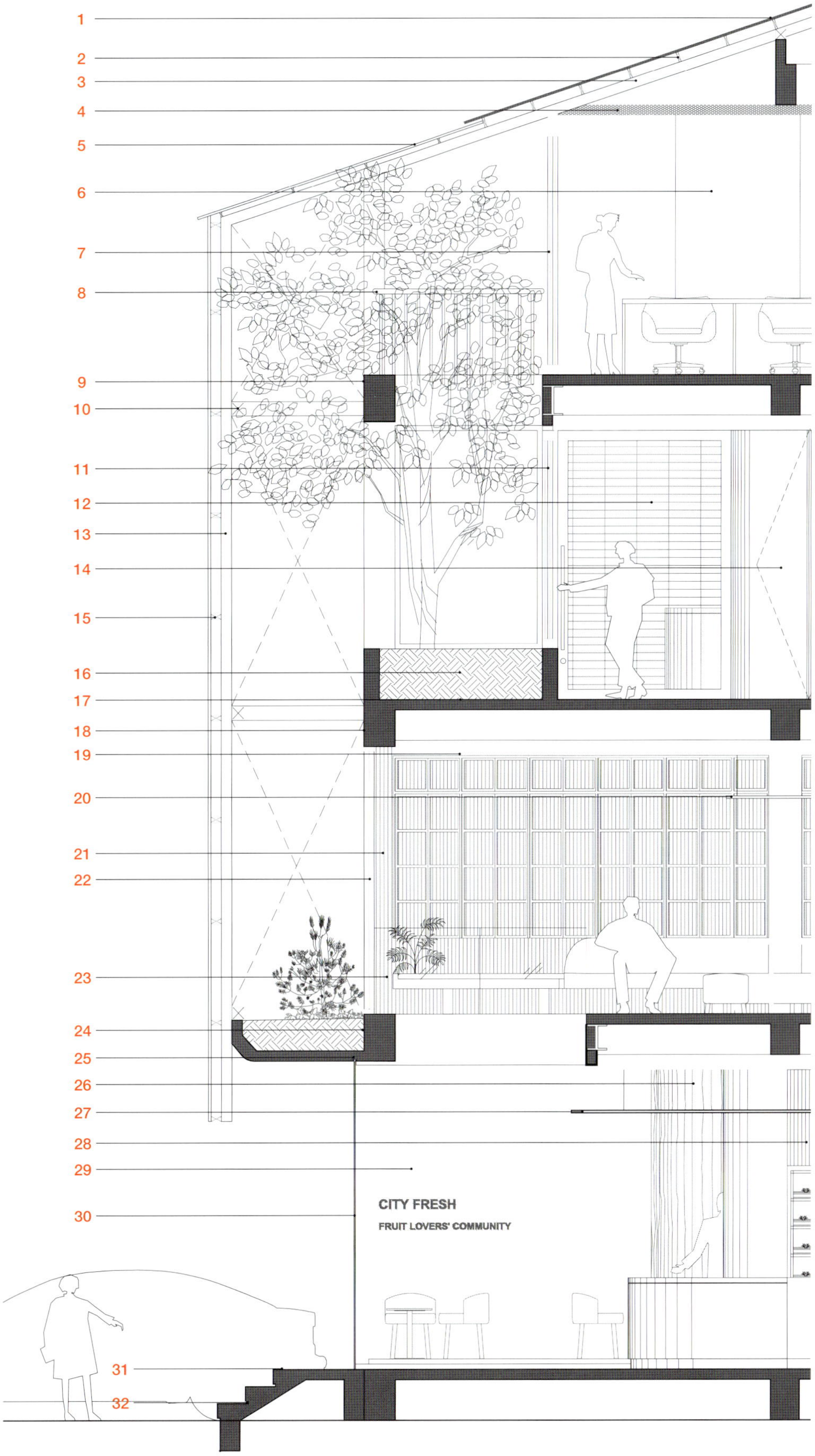

WALL SECTION DETAIL

↑ 1st floor restaurant bar & store   ← 1st floor restaurant bar   → 1st floor restaurant bar & store   ↙ Store display area   ↳ Restaurant counter

↱ 2nd floor Landscape area
→ 2nd floor group seating
→ 2nd floor tables and seating

↖ 3rd floor main kitchen　↑ 3rd floor private dining space　← Lucky tree located in 3rd floor　→ 3rd floor movable kitchenette

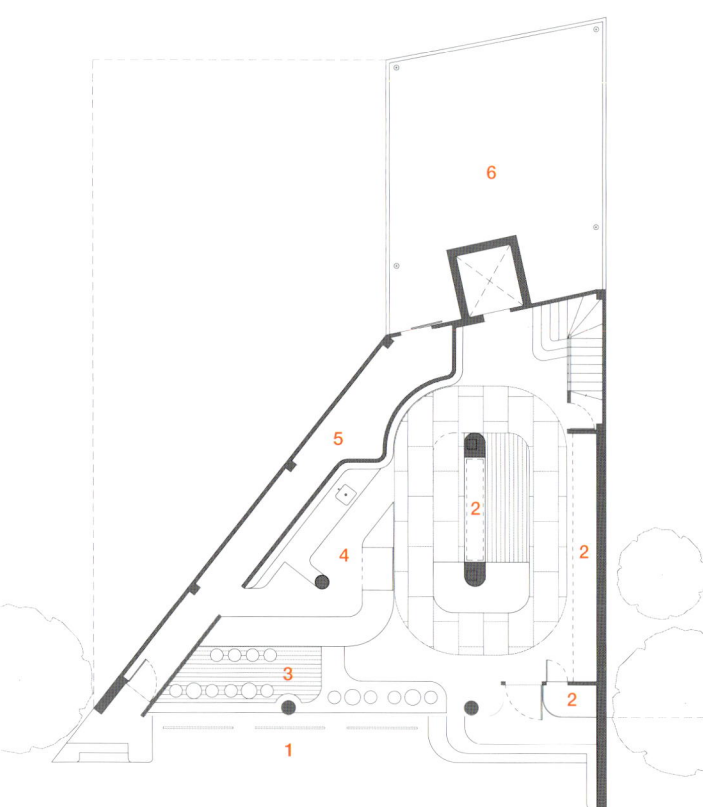
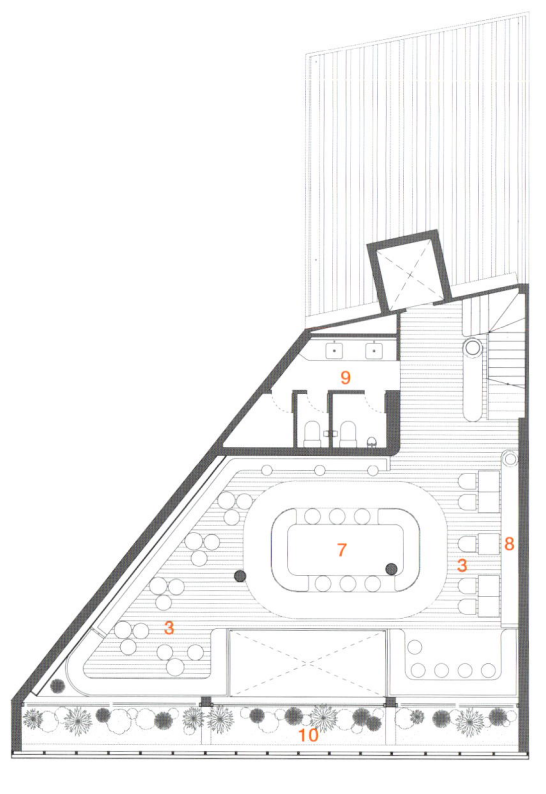

1ST FLOOR PLAN　　　　　　　　　　　　　　　　**2ND FLOOR PLAN**

← 4th floor balcony   → 4th floor working area   ↵ Toilet   ↳ Stair & seating area

1 PARKING	5 SERVICE	9 TOILET	13 COOKING STUDIO
2 DISPLAY	6 STORAGE	10 LANDSCAPE AREA	14 WAITING AREA
3 SEATINGS	7 GROUP SEATING	11 MAIN KITCHEN	15 WORKING AREA
4 COUNTER & WORK BAR	8 DINING TABLE	12 PRIVATE DINING	16 MEETING ROOM

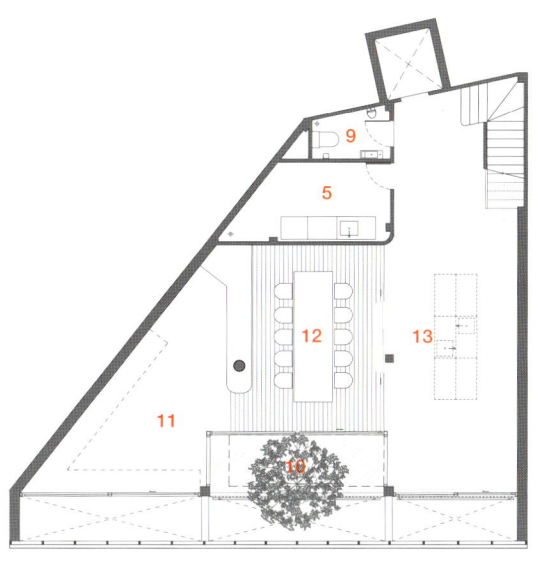

3RD FLOOR PLAN

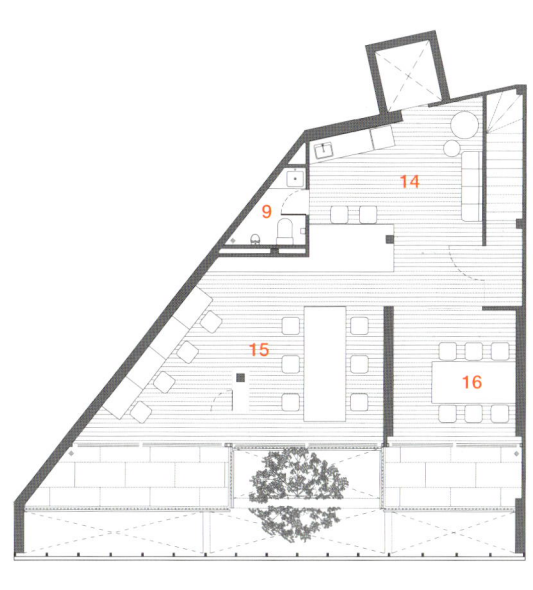

4TH FLOOR PLAN

네 개의 지붕 파빌리온
FOUR-ROOF PAVILION

ARCHITECT : FOUND PROJECTS + SCHNEIDER & LUESCHER / MIAOJIE TED ZHANG, ANTON SCHNEIDER

INSPIRED BY THE LANDSCAPE DESIGN CONCEPT OF "FOREST SCHOOL", the Four-Roof Pavilion is intended to be embedded and merged into the park. Much like the greenhouse typology, the Four-Roof Pavilion is a building of the park.

Located at the north corner of Pingshan Children's Park in Shenzhen, the Pavilion serves as a secondary gateway to the Children's Park. The 3-story, 1,200m² structure provides a pedestrian passageway, multi-purpose rooms, bookstore, cafe, & roof garden. A passageway on the ground floor allows visitors to enter via a single-story space leading them to a double-height space with an oculus skylight above. The "press and release" sequence serves as a gateway to the park. The void not only lets visitors pass through the building, but also seconds as a social corridor for seating & conversation.

A simple cast-in-place concrete structural grid is designed with two steel roofs on the south and north facades and two concrete roofs on the outdoor garden. The color palette loosely references 11th-century Chinese landscape painting, unifying the columns, beams, & diagonal bracing. The two sculptural red stairs are embedded within the structural grid, visibly highlighting the vertical circulation.

The massive roofs define the architectural identity while providing generous shading and covered space for the local community. Levels of transparency are created by the strategic placement of clear polycarbonate panels, perforated metal panels, & solid metal panels. The diamond and triangle pattern filter the natural light while introducing ambient light into the interior space. The architectural elements blur the building boundary and extend the architectural impact on both the exterior and interior, merging the relationship between architecture, landscape, and people.

Location Shenzhen, China **Use** Cultural, Public, Retail **Site area** 600m² **Built area** 1,500m² **Gross area** 1,500m² **Completion** 2022 **Project manager** Miaojie Ted Zhang **Design team** Miaojie Ted Zhang, Anton Schneider, Andri Luescher, Ryan Nguyen **Contractor** Shenzhen Sincere Environmental Art Engineering Co. Ltd., Guangdong Chengji Ecological Technology Co. Ltd. **Photographer** Schran Image

"숲속 학교"의 조경 디자인 컨셉에서 영감을 받아 4개 지붕 파빌리온을 공원에 내장하고 통합할 예정이다. 온실 유형과 마찬가지로 파빌리온은 공원의 건축물이다.

심천 핑산 어린이 공원의 북쪽 구석에 위치한 이 파빌리온은 어린이 공원으로 가는 보조 게이트웨이 역할을 한다. 3층 규모의 1,200㎡ 구조물은 보행자 통로, 다목적실, 서점, 카페, 옥상 정원을 제공한다. 1층에 있는 통로를 통해 방문객들은 단층 공간을 지나 오큘러스 채광창이 있는 이중 높이 공간으로 안내된다. 이 "누르기와 해제" 단계는 공원으로 향하는 관문 역할을 한다. 이 공간은 방문객이 건물을 통과할 수 있게 할 뿐만 아니라 앉아서 대화를 나눌 수 있는 사회적 통로로도 사용된다.

단순한 현장 타설 콘크리트 구조 격자는 남쪽과 북쪽 파사드에 두 개의 강철 지붕과 야외 정원에 두 개의 콘크리트 지붕으로 설계되었다. 색상 팔레트는 11세기 중국 풍경화를 대략 참조하여, 기둥, 보, 그리고 대각선 버팀대를 통합하였다. 두 개의 조각 같은 빨간색 계단이 구조 격자 내에 내장되어 있으며, 수직 순환을 눈에 띄게 강조하였다.

대형 지붕은 건축적 정체성을 정의하는 동시에 지역 사회에 넉넉한 그늘과 지붕이 있는 공간을 제공한다. 청정 폴리카보네이트 패널, 천공 금속 패널, 그리고 견고한 금속 패널의 전략적 배치로 다양한 투명도의 레벨을 창출한다. 다이아몬드와 삼각형 패턴은 자연광을 걸러내면서 실내 공간에 주변광을 끌어들인다. 건축 요소는 건물의 경계를 흐리게 하여 외부뿐만 아니라 내부에 걸쳐 건축적 영향을 확장하고, 건축, 경관, 그리고 사람들 사이의 관계를 융합시킨다.

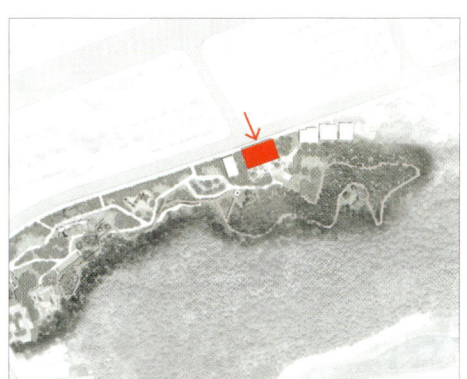

SITE PLAN

← Main view

↑ Exterior in the park view　↙ Concrete roof & cafe

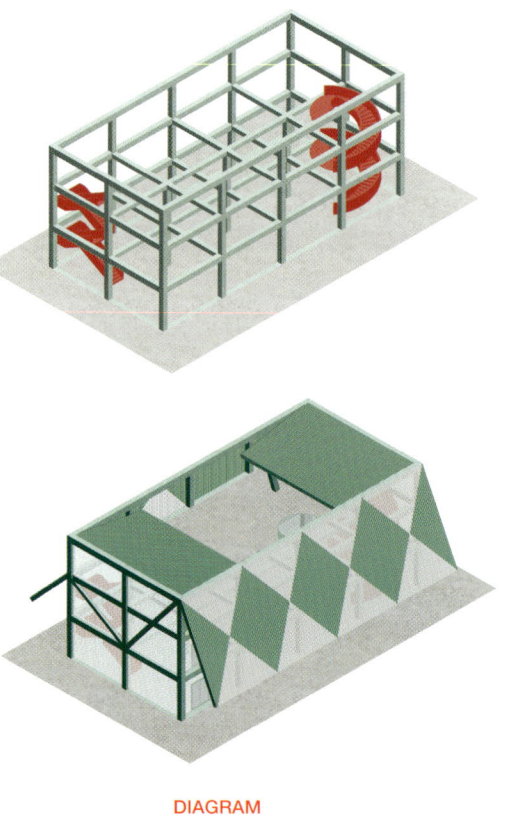

DIAGRAM

↑ Panorama view

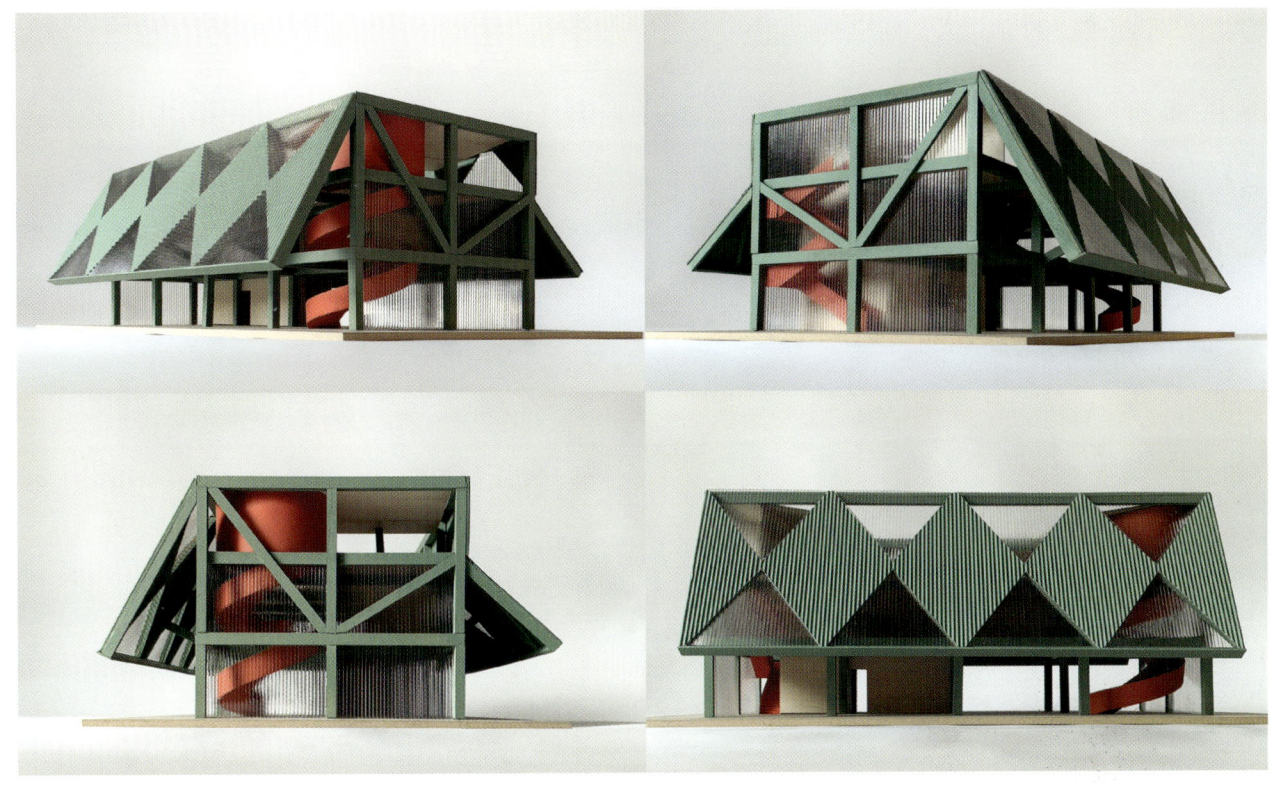

MODELLING

← Concrete roof & cafe   → Concrete roof & cafe   ↙ Diamond and triangle pattern filter the natural light

↑ Two concrete roofs on the outdoor garden   ↳ Secondary gateway to the Children's Park

103

↑ Pedestrian passageway   ↙ The oculus skylight   ↳ Sculptural red stairs

↑ Secondary gateway to the Children's Park   ↙ Roof detail, clear polycarbonate & perforated metal panels   ↘ Roof detail, clear polycarbonate & perforated metal panels

↑ 2nd floor bookstore   ↵ Sculptural red stairs

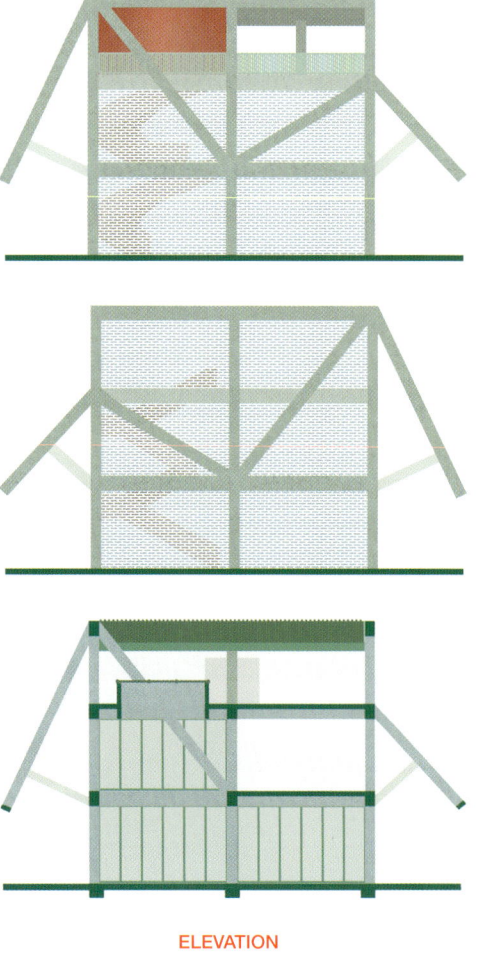

ELEVATION

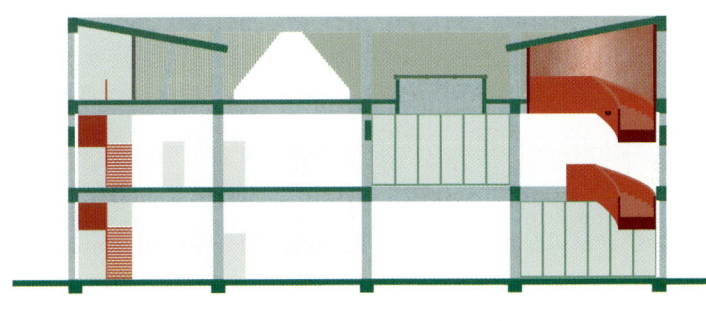

ELEVATION

1 ENTRANCE
2 TOILET
3 STAIR
4 OCULUS
5 ROOF GARDEN

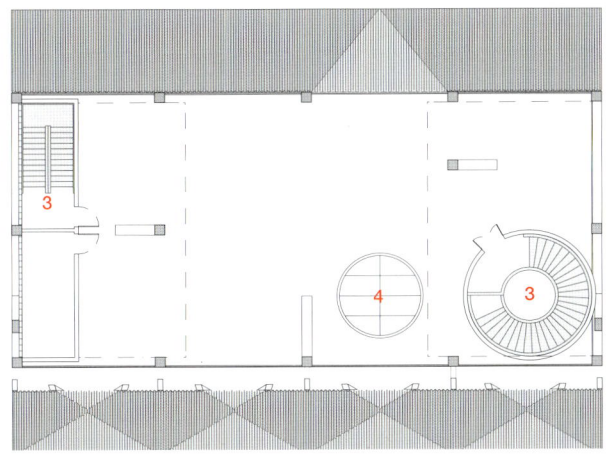

3RD FLOOR PLAN

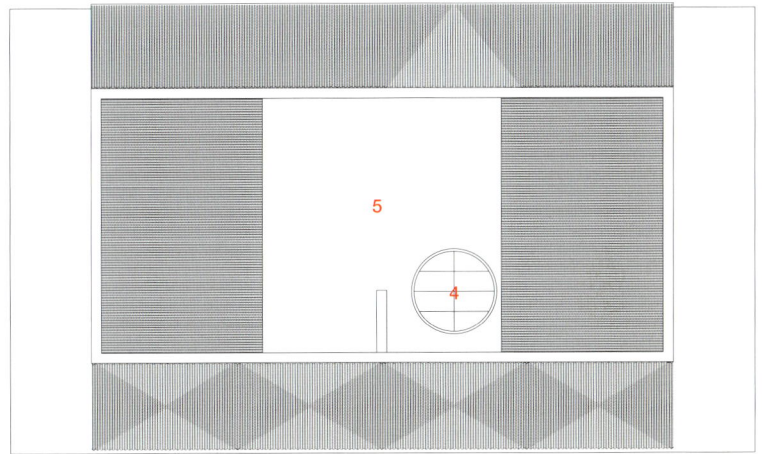

ROOF PLAN

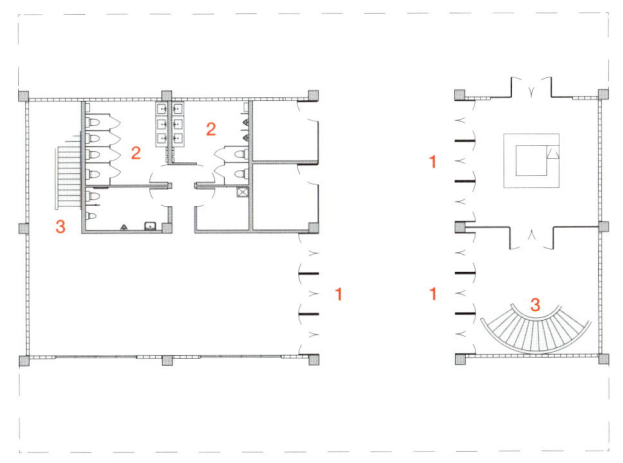

1ST FLOOR PLAN

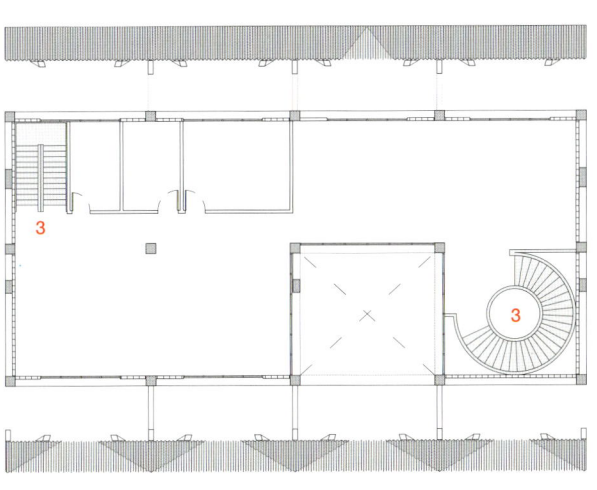

2ND FLOOR PLAN

107

MO288

ARCHITECT : HGR ARQUITECTOS / MARCOS HAGERMAN

MO288 IS A 6-STORY RESIDENTIAL BUILDING WITH 15 APARTMENTS located at Calzada Melchor Ocampo #288 in the Cuauhtemoc neighborhood in Mexico City. The project is planned on a 292m² triangular-shaped plot in a busy corner of CDMX. The objective of the project was to try to have regular spaces within a rather small and irregularly shaped piece of land. For this, 3 1-bedroom apartments between 70 and 75m² per floor were accommodated.

Something fundamental in the design was that all the spaces were well ventilated and illuminated, but trying not to have large openings due to the noise of the road that adjoins the building. The facades of the building were designed in a modular way based on brown concrete walls, ensuring that all interior spaces have open views of the city. To give more play to the facade, alternate balconies were designed, which help to give greater amplitude to the interior spaces and in turn give more privacy through metal plate railings.

At the center of the building, a patio was left to meet the required free area, which works for ventilation and lighting of the services and accommodates both vertical and horizontal circulation. All the walls of the building are load-bearing, thus avoiding having columns that generate irregular corners and thus obtaining cleaner and more regular spaces. On the ground floor of the building there is a lobby for the apartments and 2 commercial spaces that helps integrate the building with the neighborhood.

Location Colonia Cuauhtemoc, Alcaldia Cuauhtemoc, CDMX. C.P. 06500. Mexico **Use** Mixed Use **Site area** 292m² **Gross floor area** 1,398m² **Completion** 2023 **Project manager** Marcos Hagerman **Design team** Diego Castaneda, Rodrigo Duran **Development** Ciudad Vertical **Executive drawing** Rodrigo Duran **Structural design** Mata Y Triana Ingenieros Consultores **Installation** zMP Instalaciones **Photographer** Diana Arnau

MO288은 멕시코시티 쿠아우테모크 지역의 칼사다 멜초르 오캄포 288번지에 위치한 15개의 주거유닛을 갖춘 6층 규모의 주거용 건물이다. 이 프로젝트는 멕시코 시의 번화한 한편에 위치한 292㎡의 삼각형 모양의 부지에 계획되었다. 프로젝트의 목표는 상대적으로 작고 불규칙한 모양의 대지에서 규칙적인 공간을 만드는 것이었다. 이를 위해 각 층마다 70~75㎡ 규모의 한 개의 침실을 가진 아파트 세 유닛이 배치되었다.

디자인에서 중요한 것은 모든 공간이 빛이 잘 들고 환기가 잘 되는 것이지만, 건물에 인접한 도로의 소음을 고려하여 큰 개구부를 만들지 않으려고 노력했다. 건물의 파사드는 갈색 콘크리트 벽을 기반으로 모듈식으로 설계되어 모든 내부 공간이 도시의 개방된 전망을 가질 수 있도록 했다. 파사드에 더 많은 연출을 주기 위해 번갈아 설계된 발코니가 있는데, 이는 내부 공간의 광대함을 더하고 금속판 난간을 통해 프라이빗함을 갖추도록 돕는다.

건물 중앙에는 필요한 여유 공간을 충족하기 위해 중정을 두었으며, 이는 환기 및 조도에 작용하여 수직 및 수평 순환을 모두 수용한다. 건물의 모든 벽은 하중을 지지해주며 불규칙한 모서리를 생성하는 기둥을 줄여 보다 깔끔하고 규칙적인 공간을 확보할 수 있었다. 건물의 1층에는 아파트 로비와 주변지역, 건축물을 통합하는데 기여하는 2개의 상업 공간이 자리한다.

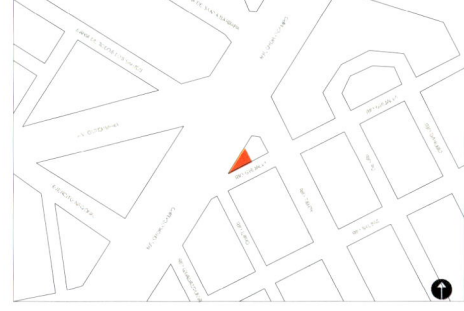

SITE PLAN

← View from northwest   → Night view

↑ View from west   ↙ Night view   → Night view

↑ Exterior view

LONGITUDINAL SECTION　　　　　　　　　　　　　　　　　　　　CROSS SECTION

111

← Patio ↑ Patio ↳ Patio ↙ Street view

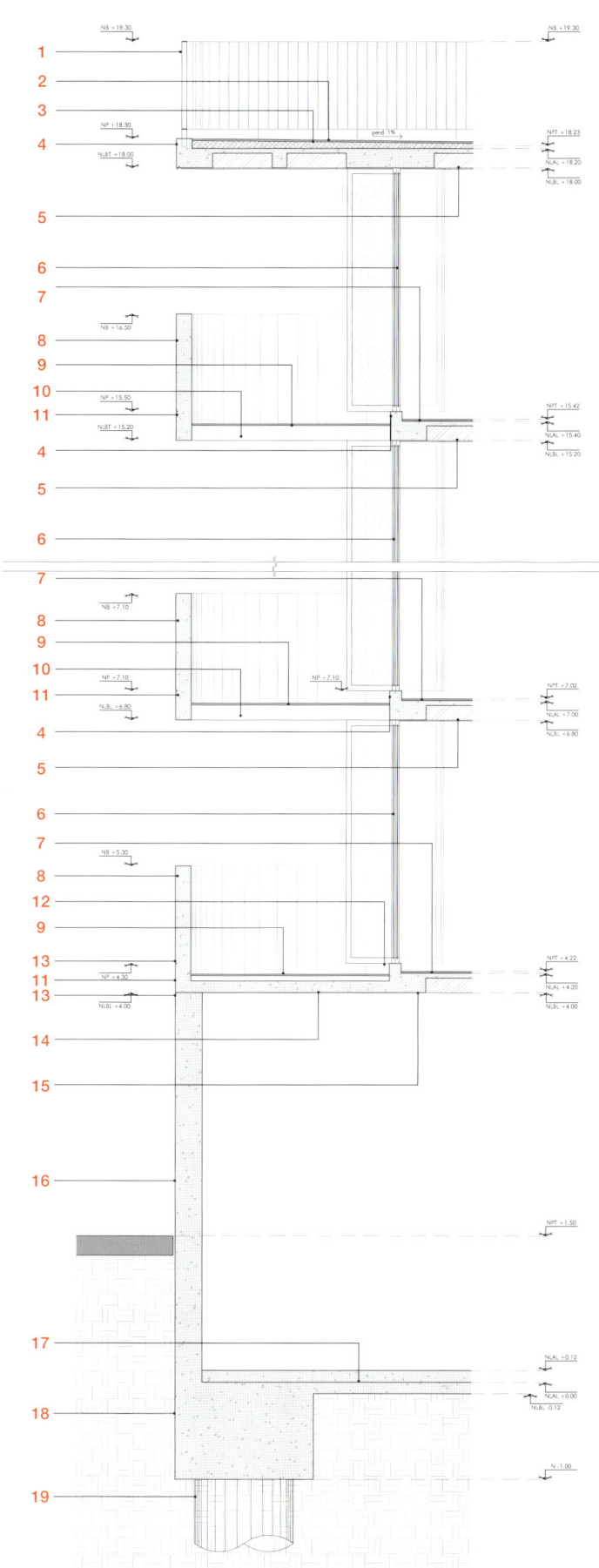

1. RAILING BASED ON A 2" STEEL SHEET FINISHED WITH ANTI-CORROSIVE PRIMER AND AUTOMOTIVE LACQUER IN MATTE BLACK COLOR S.M.A.
2. ENCLOSURE FLOOR 60×40 E=2cm. CLOSED PORE WITH MAPSA BRAND NATURAL/WET SEALANT OR SIMILAR, ADHERED WITH SAND CEMENT MORTAR.
3. SIKA BRAND WATERPROOFING MEMBRANE MODEL, SIKA MESH WITH ELASTIC, WATERPROOF AND THERMAL INSULATING COATING, SIKA ACRYLIC BRAND, ECO ROOF
4. EDGE OF REINFORCED CONCRETE SLAB AND PARABLE WITH APPARENT FINISH, FINISHED WITH APPLICATION OF SEALANT AND TWO COATS OF PI
5. PERIMETERALLY SUPPORTED SLAB WITH TOTAL DEPTH H=20cm. LIGHTENED WITH POLYSTYRENE BLOCK (SEE PLAN E-08). LOW BED FLATTENED WITH PLASTER AND FINISHED WITH 5X1 SEALER AND COMEX REALFLEX BRAND VINYL PAINT, MATTE WHITE COLOR A, TWO COATS
6. V-10 SLIDING DOOR MADE OF ALUMINUM PROFILE, 3" SPANISH SERIES PANORAMA IN DURO-E S.M.A. FINISH WITH 9mm CLEAR GLASS
7. CAPPA BRAND LAMINATE FLOOR MOD. SYNC CHROME LEYSIN OAK 8mm. BEVELED
8. 15cm PIGMENTED CONCRETE WALL. THICK (SEE PLAN E-07) DUELED FINISH, APPARENT MANZANILLO COLOR.
9. SLOTTED IPE WOOD DECK 14cm. VARIABLE WIDTH X LENGTHS WITH WEATHERPROOF TREATMENT
10. 6X4" PTR-BASED STEEL VM-1 BEAM, FINISHED WITH ANTI-CORROSIVE PRIMER AND LACQUER, AUTOMOTIVE MATTE BLACK COLOR S.M.A.
11. 15cm GRAY REINFORCED CONCRETE WALL. (SEE STRUCTURAL PLANS) APPARENT FINISH FINISHED WITH APPLICATION OF SEALANT AND TWO COATS OF MATTE WHITE VINYL PAINT, TWO COATS
12. CONCRETE SARDINEL OF 12X10cm. FINISHED, APPARENT
13. BETWEEN STREET
14. SOLID SLAB OF TOTAL CANT H= 12cm. FINISH ON ITS LOWER BED TO BE DEFINED BY TENANT.
15. PERIMETERALLY SUPPORTED SLAB OF TOTAL CANT H=20cm. LIGHTENED WITH POLYSTYRENE BLOCK FINISHED ON ITS LOWER BED TO BE DEFINED BY TENANT
16. 20cm REINFORCED CONCRETE WALL. OF THICKNESS. APPARENT FINISH TO THE EXTERIOR WITH PINE TRIPLAY FRAMING ACCORDING TO EXPLODED AND APPLICATION OF OXFORD GRAY PAINT. INSIDE WITHOUT FINISHING (TO BE DEFINED BY TENANT)
17. FIRM REINFORCED WITH TOTAL CANT 12cm, APPARENT FINISH
18. D-9 REINFORCED CONCRETE DIE
19. REINFORCED CONCRETE PILE Ø1m

DETAIL

↱ Dining & Kitchen
← Kitchen & Living room
↵ Dining room

↑ Bedroom

1 COMMERCIAL SPACE
2 DWELLING ENTRANCE
3 PARKING LOT
4 BUILDING OFFICE
5 LOBBY
6 COURTYARD
7 DWELLING UNIT
8 ROOFTOP AREA

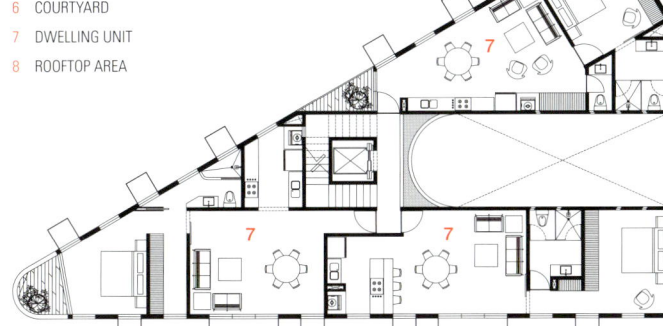

STANDARD FLOOR PLAN

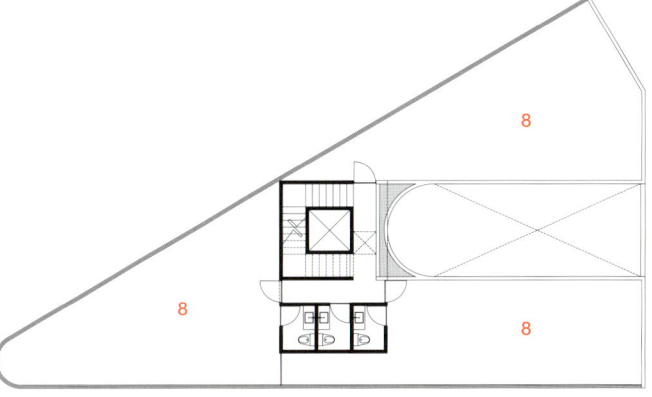

ROOFTOP FLOOR PLAN

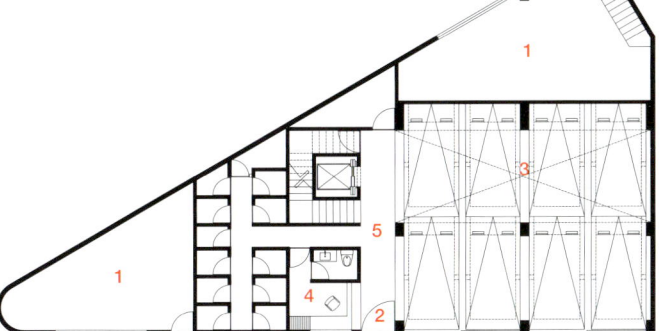

GROUND FLOOR PLAN

1ST FLOOR PLAN

테루복의 주택
HOUSE AT TERUBOK

ARCHITECT : CDG ARCHITECTS / EUGENE SEOW

THE LANDED HOUSE TYPOLOGY IS OFTEN A BATTLEGROUND OF COMPETING FACETS of public regulations, private economics, and creative design in land scarce Singapore. One common type is the semi-detached house which is a dwelling that shares a boundary party wall with its neighbour, in effect allowing only a three-sided frontage to its exterior, which more often than not is just a few metres away from the neighbour's.

House at Terubok is an architectural response to the clients' brief for a small multi-generational family house that could allow its members and other family and friends to share and enjoy ample common areas without impeding on their privacy. This was approached as an exercise of balance that maximised the permissable building envelope and expressing this maximum volume as an outer "skin", while crafting out as much quality space as possible within its compact site. Two fairface reinforced concrete walls serve as strong flanks that contain the spatial programs, while porosity and privacy at the street-front are balanced using facade screening and greenery, which presents the front of the house organically, as a contrast (and complement) to the strict and controlled demeanour of its sides. Planters within the naturally ventilated incision between the bathrooms and party wall provide an environmental filter for cross-ventilation, and in turn help create a meaningful connection to the outdoor space that the existing site lacked.

The interior layouts are conceptualised around the creation of an airwell that captured as much natural light and air as possible, and by pulling the main living spaces and bedrooms away from the party wall, creates a naturally ventilated incision that runs high and long. This strategy of organising as much of the interior spaces around this airwell and internalising the aspect of these spaces, focuses on taking advantage of the few things that are still abundantly free yet very important in a small site ? light and air, while still maintaining privacy from the neighbours.

땅이 부족한 싱가포르에서 주택 유형은 종종 공공 규제, 사적 경제성, 창의적인 디자인이 경쟁하는 전쟁터와 다름없다. 한 가지 일반적인 유형은 이웃과 경계벽을 공유하는 주택인 반단독주택으로, 사실상 외부에 3면만 노출되는 경우가 많으며 이웃집과 불과 몇 미터 떨어져 있는 경우가 많다.

테루복의 주택은 가족 구성원과 다른 가족 및 친구들이 프라이버시를 침해하지 않으면서도 넓은 공용 공간을 공유하고 즐길 수 있는 소규모 다세대 가족 주택을 원하는 클라이언트의 요구에 대한 건축적인 응답이라고 할 수 있다. 이 프로젝트는 허용되는 건물 외피를 최대화하고 이 최대 부피를 '외피'로 표현하는 동시에 좁은 대지 내에서 가능한 한 많은 양질의 공간을 만드는 균형 잡힌 접근 방식으로 진행되었다. 두 개의 치장 철근 콘크리트 벽은 공간 프로그램을 담는 강력한 측면 역할을 한다. 도로변의 다공성과 프라이버시는 엄격하고 통제된 측면의 태도와 대비를 이루며 유기적으로 집의 전면을 드러내는 파사드 스크린과 녹지를 통해 균형을 맞췄다. 욕실과 경계벽 사이의 자연 환기가 가능한 절개부 내 화분은 교차 환기를 위한 환경 필터 역할을 하며, 기존 부지에 부족했던 야외 공간과의 의미 있는 연결성을 만들어낸다.

내부 레이아웃은 자연 채광과 공기를 최대한 끌어들이는 통풍구를 만들고, 주요 거실 공간과 침실을 경계벽에서 멀리 떨어뜨려 자연 환기가 가능한 높고 긴 절개부를 만드는 것을 콘셉트로 삼았다. 이 통풍공간을 중심으로 내부 공간을 최대한 많이 구성하고 이러한 공간의 양상을 내재화하는 이 전략은 작은 대지에서 여전히 풍부하면서도 매우 중요한 몇 가지 요소, 즉 빛과 공기를 활용하면서도 이웃과의 프라이버시를 유지하는 데 중점을 둔다.

Location Terubok, Singapore **Use** Housing **Site area** 228m² **Building area** 380m² **Gross floor area** 317m² **Completion** 2022 **Project manager** RJ Zheng **Contractor** Strategic Engrg and Construction Pte Ltd **Photographer** Ong Chan Hao

CONCEPT DIAGRAM

← Front facade

↑ Exterior view   ↓ Partly view of exterior

← Night view → Entrance

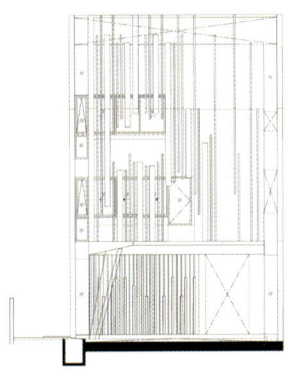

FRONT ELEVATION

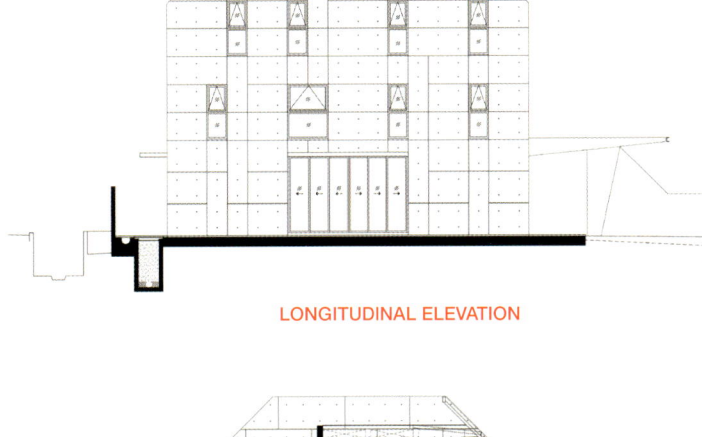

LONGITUDINAL ELEVATION

CROSS SECTION

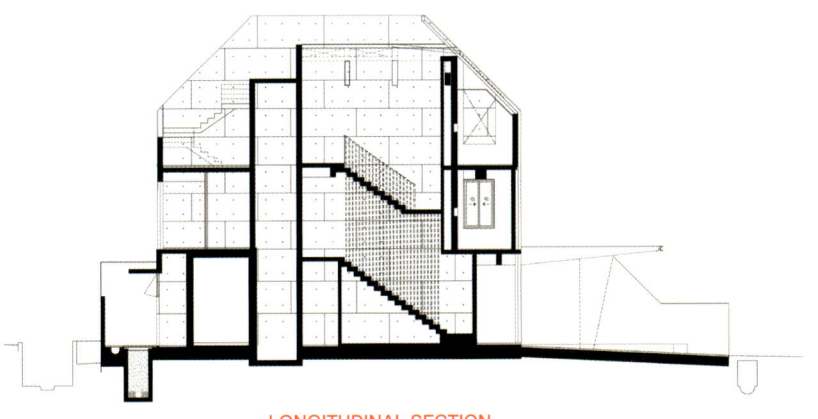

LONGITUDINAL SECTION

1. 5mm THK STEEL STEEL PROFILE WITH LIGHT GREY MATT POWDER COAT FINISH
2. 33.7mm DIA X 3mm THK STEEL ROD, POWDER-COATED. FIXING AND SIZE TO C&S EGNR'S DETAILS.
3. ROD WELDED TO L PLATE
4. SCREED WITH EPOXY FINISH
5. ROD WELDED TO STEEL SECTION
6. HEAT-STRENGTHENED LAMINATED LOW-E GLASS PANEL
7. SKYLIGHT STEEL SUPPORT STRUCTURE TO PE DETAILS
8. DOUBLE BANK WEATHERPROOF LOURVE
9. CALCIUM SILICATE BOARD SKYLIGHT SUNSHADE

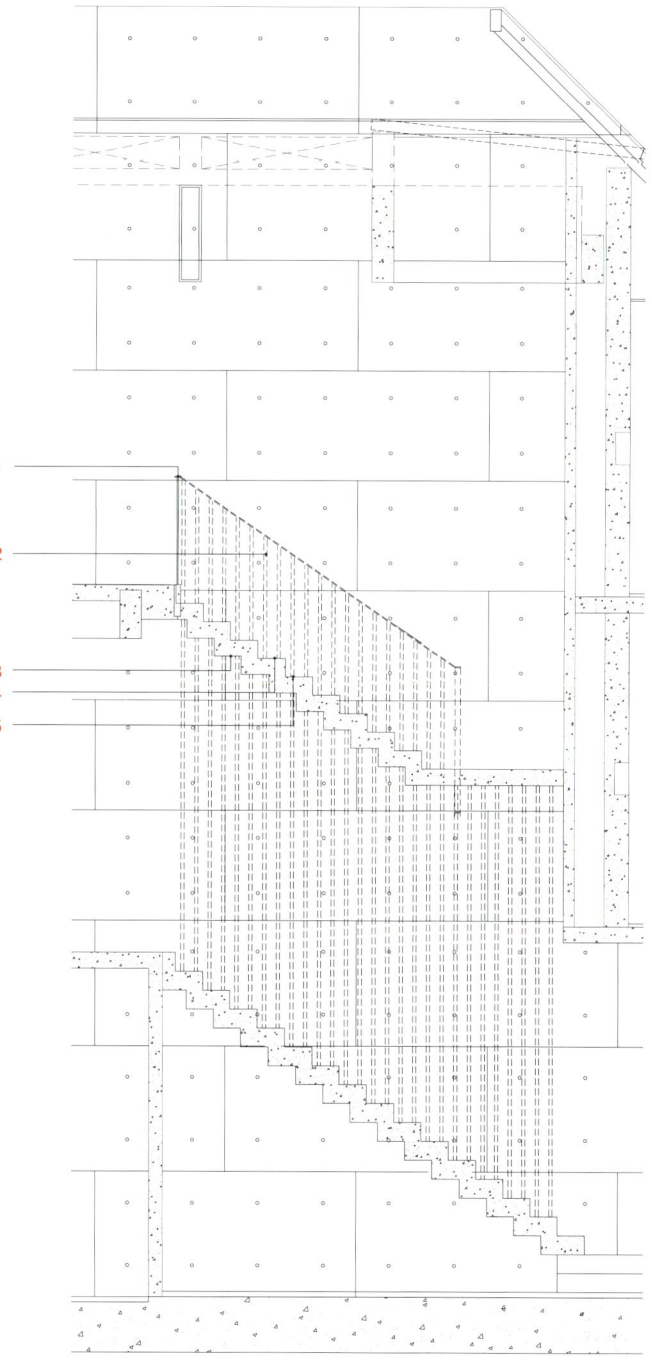

Courtyard

STAIRCASE LONG SECTIONAL DETAILS

↖ Stair & Courtyard   ↙ Skylight   ↑ Corridor

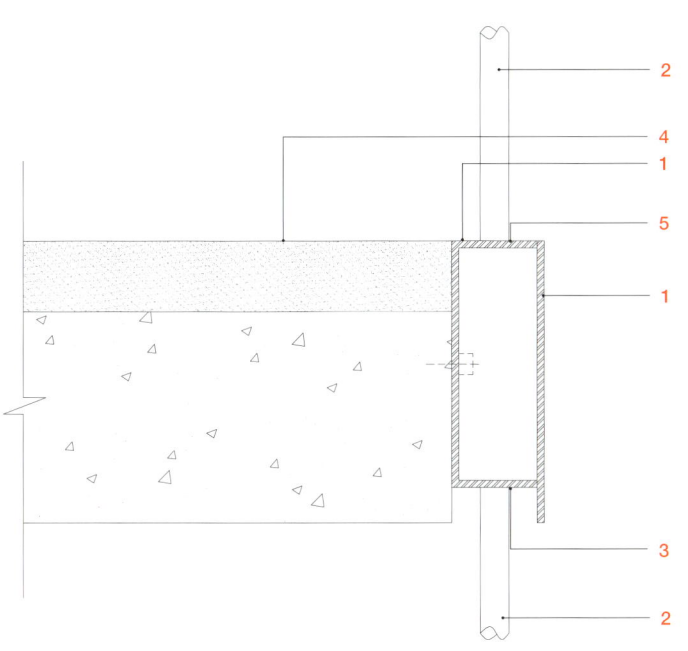

STAIRCASE SHORT SECTIONAL DETAILS

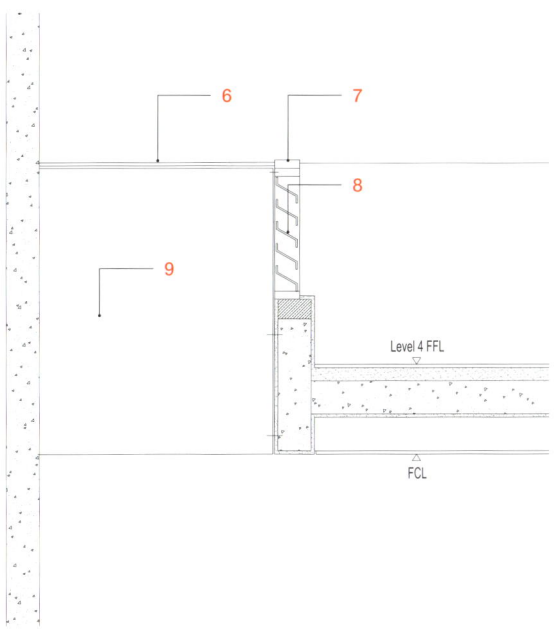

SKYLIGHT SECTIONAL DETAILS

↑ Living room   ↙ Dining room   ↳ A gap between the building and the exterior wall

↑ Bedroom

1 MAIN ENTRANCE	4 KITCHEN	7 BATHROOM	10 MASTER BEDROOM
2 LIVING ROOM	5 HELPER'S ROOM	8 BEDROOM	11 BALCONY
3 DINING ROOM	6 HOUSEHOLD SHELTER	9 STUDY	

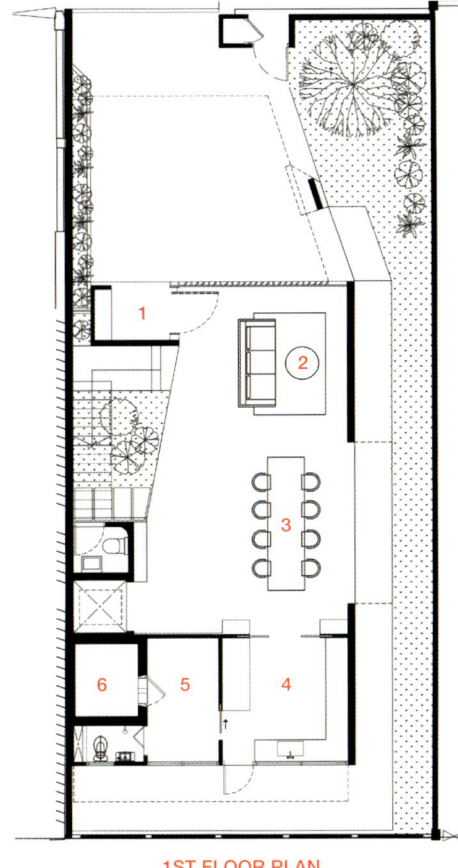

1ST FLOOR PLAN

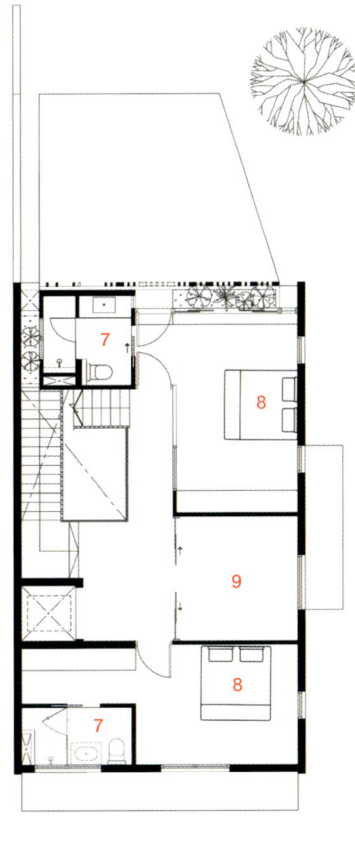

2ND FLOOR PLAN

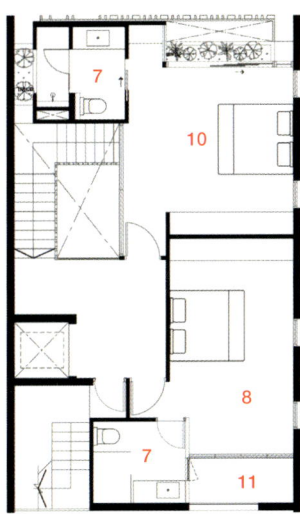

3RD FLOOR PLAN

플롭 아트 스페이스
FLOP ART SPACE

ARCHITECT : ARCHIPOETRY STUDIO / LANGJIN ZHU

FLOP IS A RENOVATION PROJECT OF AN OLD FACTORY. It is located in an old industrial zone on Jugong Road, Binjiang District, Hangzhou City. There are some industrial factories remains around it, a high-tech industrial park and creative pioneer center are constantly built out. The owner is an individual entrepreneur. She hopes that the Flop art space can meet various needs such as commercial photography, with a fusion of art and brand exhibition, buyer's shop, coffee shop and other leisure space. Flop's original building is a steel factory made of colored composite steel with decayed and damaged facades, but the structural framework is intact. Flop revitalizes the old factory building by retaining the structure of the old factory building, combining it with contemporary space needs, and responding to old building materials. we strengthened the original structure and removed the old composite steel wall and roof which had been seriously decayed. The volume and height of the renovated building will be the same as the original. Due to the multi demand of photography shooting, we added additional two mezzanine to the existing double height factory building. Both the existing and original structure supported loads independently.

In terms of spatial quality, there is a large area of empty space on the first floor, retaining only the spaces required for human circulation, entrances and exits. The second floor is added according to the functional requirements of photography space. A small device is placed on the second floor to divide the space. The roof uses solar panels to introduce indoor lighting, and the walls use aluminium alloy metal to reflect the original architecture.

The selection of facade materials corresponds to the original style of the old building. Likewise, aluminium alloy roofs are more durable, which is mostly used for walls and roofs. Glass is much simpler without a frame. We hope to simply highlight the materiality of the aluminium alloy. Through the change of architectural materials, we respond to the requirement of commercial photography space with different textures. During construction, the damp walls naturally fell down and formed a special peeling effect. This peeling effect revealed the original building materials beneath, and we kept this wall effect that was created by accident.

Location Jugong Road, Binjiang District, Hangzhou City, Zhejiang Province, China **Use** Exhibition & commercial space **Site area** 1,824m² **Built area** 1,032m² **Gross area** 807m² **Completion** August, 2021 **Project manager** Langjin Zhu **Design team** Xiuying Xiao, Zheqi Shen, Li Yu, Wenjing Wang, Wenying Xie, Xinjia Niu, Yue Qi **Contractor** Hangzhou Zhongchao construction material Pte. Ltd, Hangzhou Youban decoration construction **Photographer** Jianbo Ke

SKETCH

← Exterior view → Before renovation

backyard
这不是个展

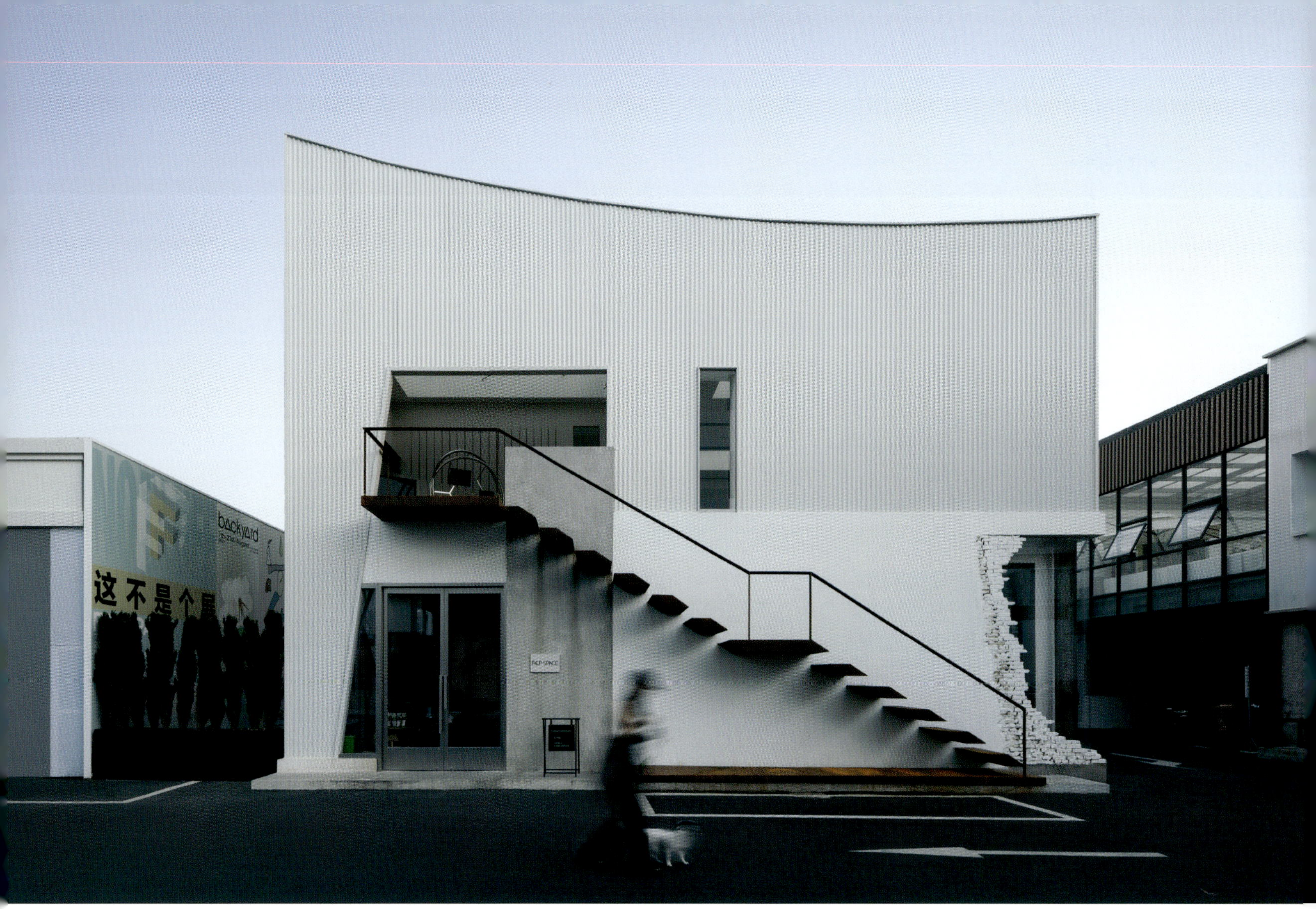

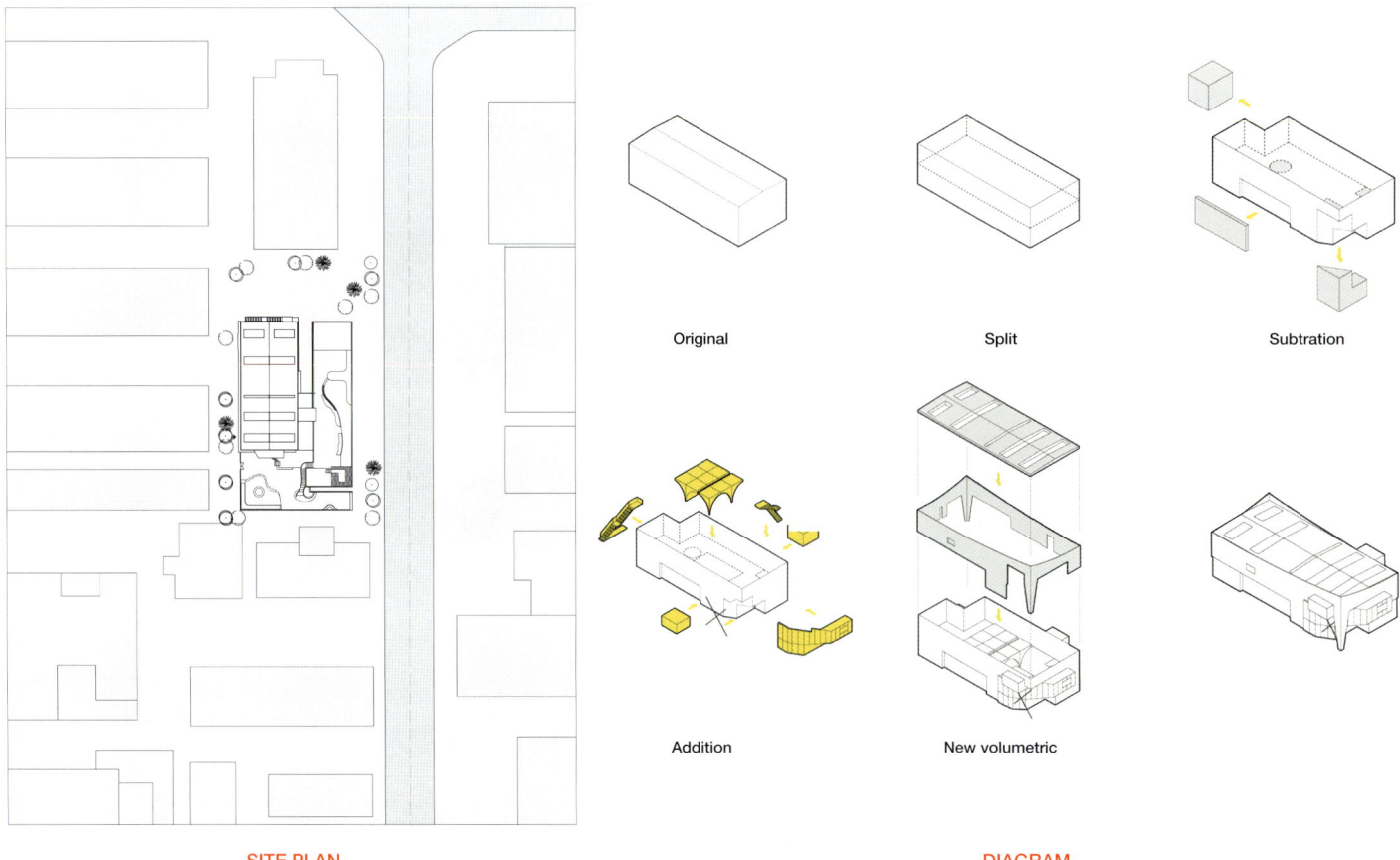

North entrance view

SITE PLAN

Original
Split
Subtration
Addition
New volumetric

DIAGRAM

↑ South entrance view

Structure before renovation

Demolition part

New-built structure

플롭은 오래된 공장을 개조하는 프로젝트이다. 항저우시 빈장구 죽공로의 옛 산업 지대에 있다. 주변에는 여전히 일부 산업 공장들이 남아 있으며, 고기술 산업단지와 창조적 선도 센터가 지속해서 건설되고 있으며 소유주는 개인 기업가이다. 그녀는 플롭 아트 공간이 상업 사진 촬영은 물론, 예술과 브랜드 전시, 바이어 숍, 커피숍 및 기타 여가 공간 등 다양한 요구를 충족시킬 수 있기를 희망하였다. 플롭의 본래 건물은 색상이 입혀진 복합강재로 이루어진 강철 공장으로, 외관은 부식되고 손상되었으나 구조적인 골격은 온전히 유지되고 있었다. 플롭은 기존 공장 건물의 구조를 보존하며 현대적 공간 요구사항과 결합하고, 노후화된 건축 자재에 대응하여 오래된 공장 건물에 새로운 생명을 불어넣었다. 우리는 기존의 구조를 강화하고, 심각하게 부식된 복합 강재 벽과 지붕을 제거하였다. 개조된 건물의 부피와 높이는 원래 건물과 동일하게 유지될 것이다. 사진 촬영의 다양한 요구를 충족시키기 위해, 우리는 기존의 이중 높이 건물에 추가적으로 두 개의 중층을 추가하였다.

기존 구조와 원래 구조 모두 각각의 하중을 독립적으로 지지하고 있다.

공간의 질적 측면에서 첫 번째 층에는 사람의 순환, 출입구 등 필수 공간만을 남기고 넓은 공간이 있다. 두 번째 층은 사진 촬영 공간의 기능적 요구에 맞추어 추가되었다. 공간을 분할하는 소형 장치가 두 번째 층에 설치되었다. 지붕은 실내조명을 도입하기 위해 태양광 패널을 사용하며, 벽은 원래 건축물을 반영하기 위해 알루미늄 합금 금속을 사용하였다. 구시대 건물의 본래 스타일을 존중하여 파사드 재료를 선정하였다. 마찬가지로, 알루미늄 합금 지붕은 내구성이 뛰어나 벽과 지붕에 주로 사용하였다. 틀이 없어 훨씬 단순한 유리 재질의 미를 강조하고자 하였다. 우리는 알루미늄 합금의 재질감을 단순하고 돋보이도록 원했다. 건축 재료의 변화를 통해, 상업 사진 공간의 요구에 다양한 질감으로 응답하고자 하였다. 건축 과정 중에 습기가 찬 벽이 자연스럽게 무너져 특별한 벗겨짐 효과를 형성하였다. 이 벗겨짐 효과는 원래 건물의 재료를 드러냈고, 우리는 이 우연한 벽효과를 보존하기로 하였다.

← Exterior corner detail view   → West window view   ↙ West exterior view

↑ South entrance & courtyard

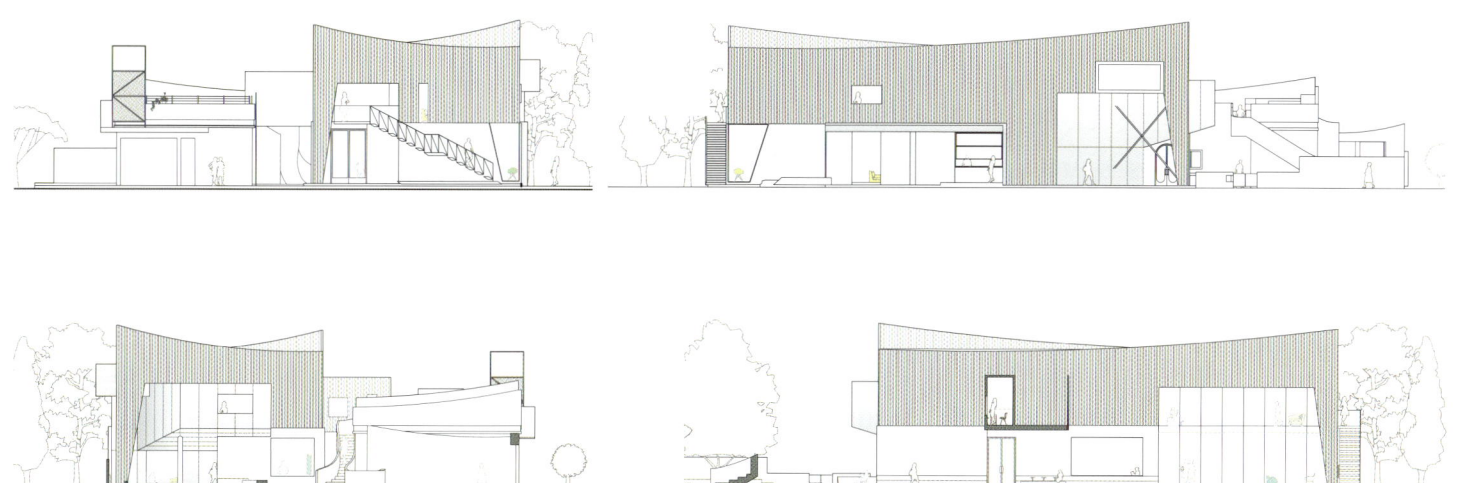

ELEVATION

↑ Courtyard top view   ↵ Exterior detail   ↓ Exterior detail   ↳ Roof top

1. MANUFACTURER'S PROCEDURE OF TEXTURE PAINT
   PUTTY THREE TIMES
   5mm 1:03:2.5 LEVELING OF CEMENT-LIME MORTAR
   13mm 1:03.3 BRUSHING CEMENT-LIME MORTAR PRIMER
   THE ORIGINAL RED BRICK WALL OF THE FACTORY
   STEEL COLUMN
   SINGLE-LAYER PRESSED METAL PLATE
   DRYWALL CEILING
2. STRUCTURE OF LIFTING EQUIPMENT FROM THE ORIGINAL FACTORY
3. 0.9mm ALUMINUM MAGNESIUM ROOF PANEL
   PERMEABLE MEMBRANE
   100mm ULTRA-FINE GLASS WITH THERMAL INSULATION FOAM
   NON-WOVEN FABRIC
   0.5mm COLOR STEEL PLATE
4. 35mm ROUGHCAST
   1:3 CEMENT MORTAR LAYER
   120mm REINFORCED CONCRETE & COMPACTED STEEL COMPOSITE FLOOR
   DRYWALL CEILING
5. 20mm TERRAZZO FINISH
   20mm 1:3 CEMENT MORTAR LAYER
   INTERFACE AGENT
   100mm C20 CONCRETE
   150mm GRADED GRAVEL
   RAMMED EARTH

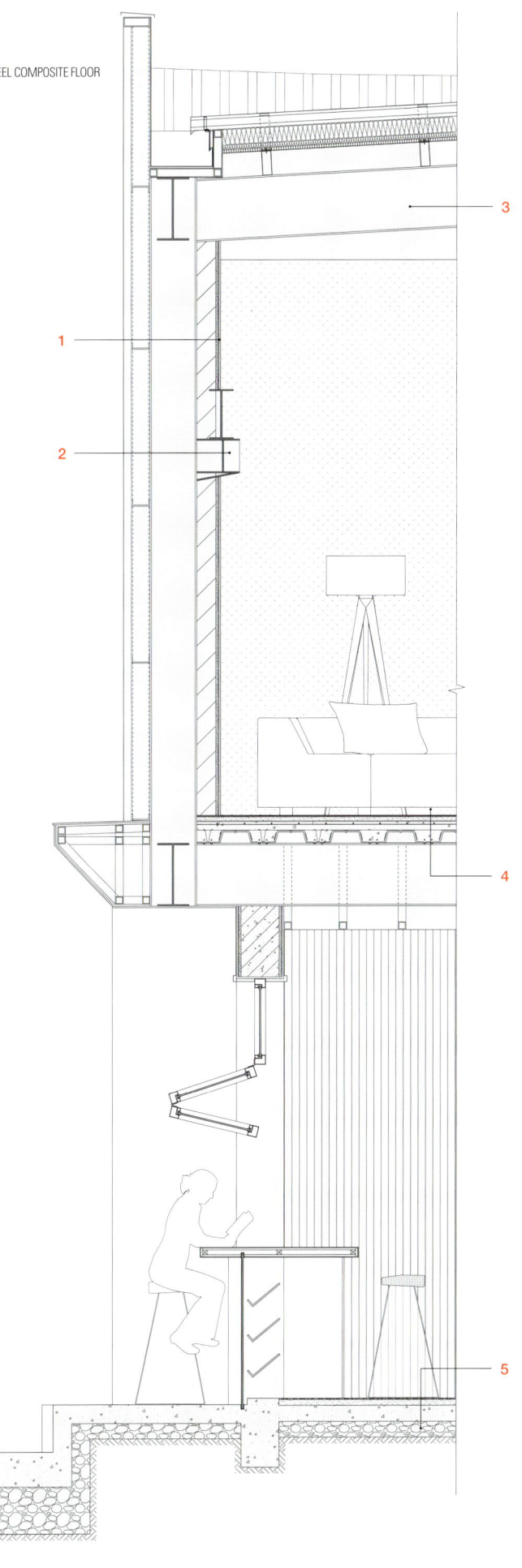

DETAIL

↑ Courtyard detail   ↓ Courtyard detail

↑ 1st floor entrance view　↙ The wall special peeling effect　↘ Stair & wall detail

↑ Discussion area
↓ North entrance

↑ 2nd floor exhibition area

1 ENTRANCE
2 BUYER'S SHOP
3 LOUNGE
4 CAFE
5 DISCUSSION AREA
6 EXHIBITION
8 READING CORNER
9 COURTYARD
10 MOVIE TERRACE
11 OUTDOOR LAWN
12 ROOFTOP BAR\
13 SKY LINK

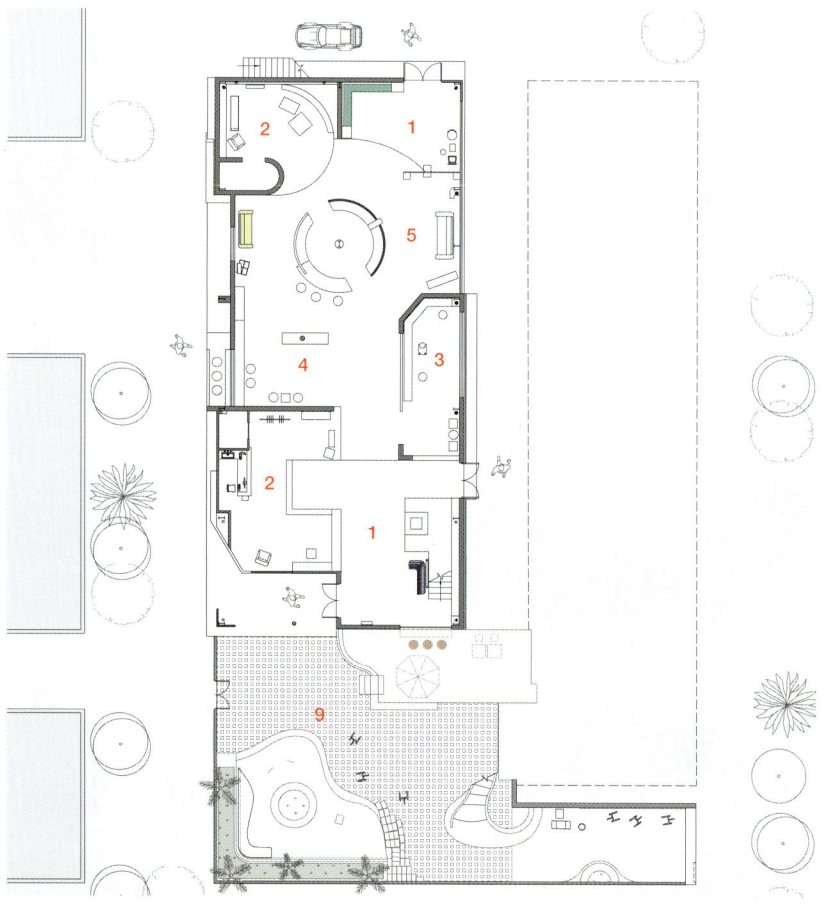

1ST FLOOR PLAN

← 2nd floor stair & wall effect → 2nd floor exhibition area ↙ 2nd floor exhibition area ↳ 2nd floor exhibition area

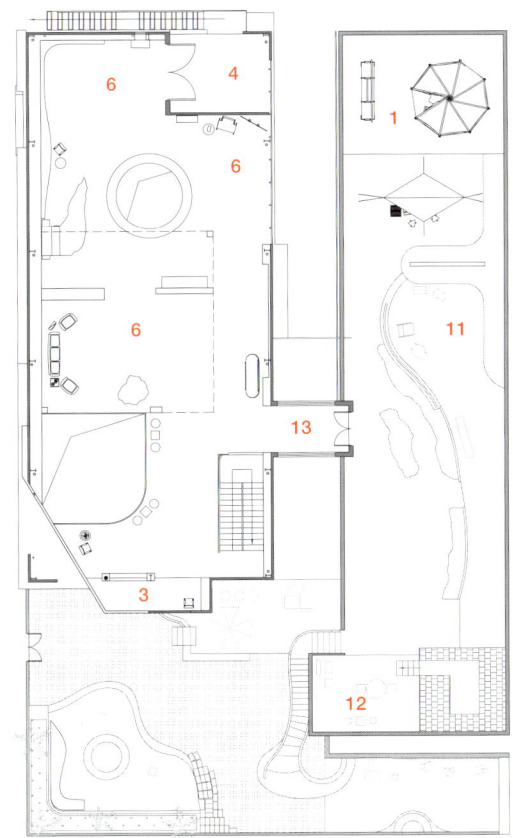
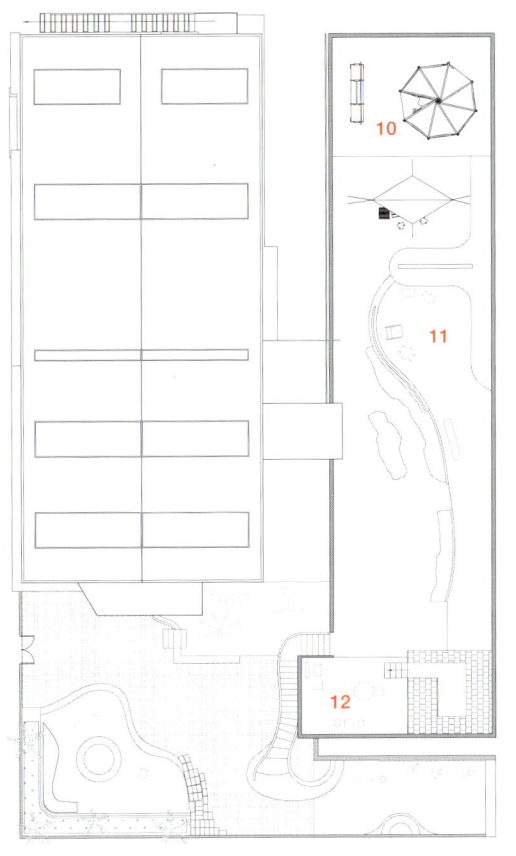

2ND FLOOR PLAN

ROOF PLAN

카레타스
CARRETAS

ARCHITECT : HERYCO / LUIS CARLOS AGUILAR GONZALEZ

CARRETAS IS A RENOVATION PROJECT OF AN APARTMENT BUILDING located in a residential, family-friendly, and pedestrian neighborhood in Queretaro, with proximity to and views of the iconic Aqueduct of Queretaro. This building, originally constructed in the 90s, now features four spacious three-bedroom apartments, each with a study, ideal for families and professionals seeking a modern and functional space. Additionally, the ground floor of the building houses an architecture office, making it even more attractive with its direct connection to the street and serving as an anchor component in the building. During the renovation of the building, we faced some unexpected challenges, such as the outdated construction system used for the building's structure. This system, based on steel beams and lightweight concrete slabs, limited our remodeling options and prevented us from demolishing partition walls. However, we managed to find an aesthetic solution by leaving the exposed steel beams uncovered, enhancing the spaciousness and natural lighting in the spaces. Despite these challenges, our goal was to give the building a more youthful image through the use of pigmented lime stucco, steel details, art, and furniture, creating a cohesive and elegant visual effect. We used "Nanocal" pigmented lime stucco for the entire facade and courtyards, and reinforced the building's structure, as the entire rooftop slab structure was corroded and required uncovering, cleaning, treating, and reinforcing from underneath.

In this renovation, we prioritized the comfort of the residents by installing an efficient heating system and ensuring optimal water pressure at all times. We also opened up terraces to enjoy the views of the aqueduct and installed high-quality furniture and mattresses to guarantee rest and comfort for the residents. To complement the modern aesthetics of the building, we integrated artwork by various Mexican artists, creating a unique and sophisticated atmosphere.

This renovation project is a great example of how architecture can transform spaces and improve people's quality of life. Moreover, this project represents a positive contribution to the city by reclaiming an abandoned building and transforming it into a habitable and functional space. The renovation of unused or abandoned buildings not only provides a solution to housing shortages but also reduces pressure on land use and prevents uncontrolled urban expansion. This project exemplifies how building renovation can have a positive impact on the city and its inhabitants.

Location Carretas, Queretaro, Mexico **Use** Accommodation & Residential **Gross floor area** 550m² **Completion** 2023 **Project manager** Luis Carlos Aguilar Gonzalez, Jose Carlos Hernandez Martinez **Photographer** Ariadna Polo Fotografia

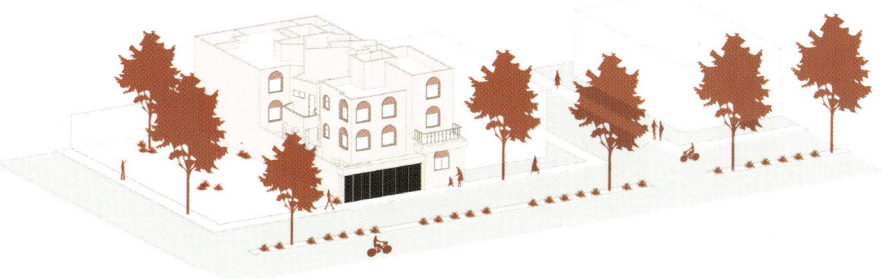

ISOMETRIC

← Front facade

← Exterior view　→ Back view

카레타스는 케레타로의 보행자 친화적이고 가족 중심의 주거 지역에 위치한 아파트 건물의 리노베이션 프로젝트로, 상징적인 케레타로 수도교의 전망과 근접성을 자랑한다. 1990년대에 지어진 본 건물은 현재 넓은 세 개의 침실과 서재를 갖춘 아파트 4채로 구성되어 있으며, 현대적이고 기능적인 공간을 찾는 가족과 전문가에게 이상적인 곳이다. 또한, 건물의 1층은 건축 사무소로 사용되며, 거리와 직접 연결되어 있어 건물의 중심 구성 요소로서 더욱 매력적이다.

리노베이션 과정에서 건물 구조에 사용된 오래된 건축 시스템과 같은 예기치 않은 어려움에 직면했다. 철골 보와 경량 콘크리트 슬래브를 기반으로 한 이 시스템은 리모델링 선택권을 제한하고 벽 철거를 방해했다. 그러나 그들은 노출된 철골 보를 그대로 둠으로써 공간의 면적과 채광을 강화하는 미학적 해결책을 찾을 수 있었다. 이러한 도전에도 불구하고, 그들의 목표는 유색 석회 스투코, 철골 디테일, 예술 및 가구를 사용하여 건물에 보다 젊은 이미지를 부여하는 것이었다. 옥상 슬래브 구조 전체가 부식되어 하부의 청소, 처리 및 보강이 필요했기에 전체 파사드와 중정에 '나노컬' 안료 석회 스투코를 사용하고 건물 구조를 강화하였다.

금번 리노베이션에서 그들은 효율적인 난방 시스템을 설치하고 상시 최적의 수압을 보장하는 등 거주자의 쾌적함을 최우선으로 고려하였다. 또한 테라스를 개방하여 수도교의 전망을 감상하고, 거주자들에게 휴식과 편안함을 보장하기 위해 고품질 가구와 매트리스를 설치하였다. 건물의 현대적인 미학을 보완하기 위해 멕시코 예술가들의 다양한 작품을 결합하여 독특하고 세련된 분위기를 조성했다.

이 리노베이션은 건축이 어떻게 공간을 변화시키고 사람들의 삶의 질을 향상시킬 수 있는지 보여주는 훌륭한 사례이다. 또한 이 프로젝트는 버려진 건물을 회수하여 거주 및 기능적인 공간으로 변화시킴으로써 도시에 긍정적인 기여를 나타냈다. 사용되지 않거나 버려진 건물의 개조는 주택 부족에 대한 해결책을 제공할 뿐만 아니라 토지 사용에 대한 압력을 줄이고 무분별한 도시 확장을 방지한다. 더불어 건물 리노베이션이 도시와 그 주민들에게 긍정적인 영향을 미칠 수 있음을 보여준다.

← Arch column → Entrance passage

FRONT ELEVATION

Passage
 Front yard

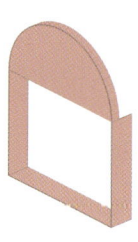

DETAIL OF BLACK WORK

← Rest area → Stucco column

1 ENTRANCE
2 COURTYARD
3 REST AREA
4 ARCHITECT'S OFFICE
5 RESIDENTIAL UNIT
6 TERRACE

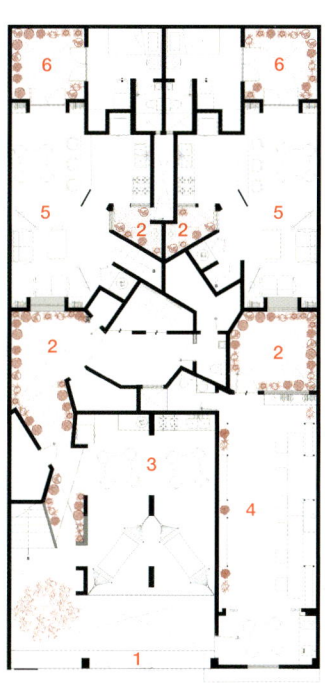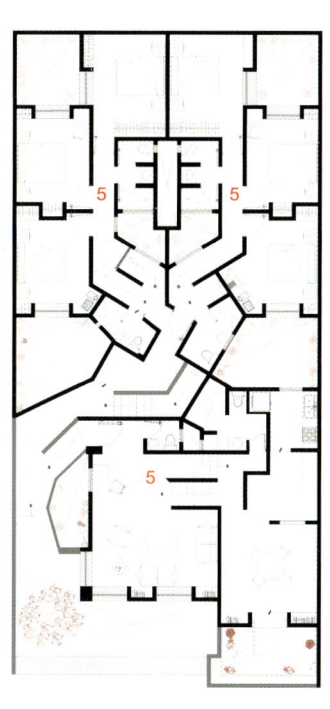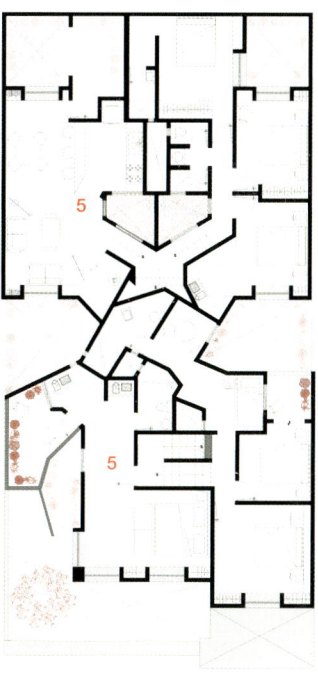

GROUND FLOOR PLAN   **1ST FLOOR PLAN**   **2ND FLOOR PLAN**

← Office → Office ↙ Terrace ↳ Front yard

↑ Office on the ground floor

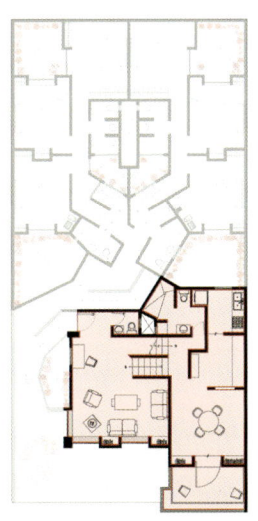

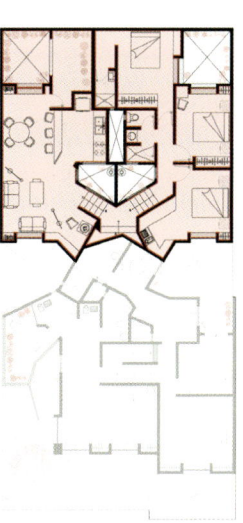

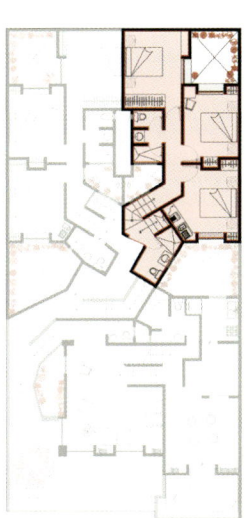

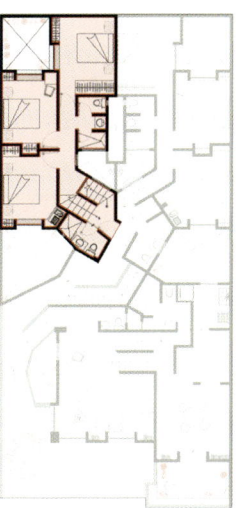

DIVISION OF EACH RESIDENTIAL UNIT

완산뜨락 주민소통방

WANSANTRAK COMMUNITY CENTER

ARCHITECT : YOAP ARCHITECTS LTD. / SANGKYONG JEONG, INKEUN RYOO, DORAN KIM

MEMORY OF THE CITY : PUBLIC GARDEN WITH LOW THRESHOLD

<Wansantrak Community Center> was designed to brighten up the corner of the alley in Wansandong and create a gathering space to share the lives of neighbors. Various architectural ideas were developed to create a comfortable space for residents. As a small public building, the main concept was to make the building more approachable and open to public. We proposed to design the alley in front along with the building. Starting with the design of the alley, the architecture is intimately intertwined with surrounding space creating a lively flow to the village.

New color was added to the old alley, and the color extends to the entrance of the building. Linear patterns recreate the image of the old alley, and implies the image of a crosswalk that makes pedestrians to pause for once when passing through.

Expansion : An Unintentional Encounter & Friendly building

We hoped that the public architecture with a low threshold would be a space that anyone can stop by. It could be a place to take shelter from rain or stop and sit having a small talk at 'toetmaru;bench'. Open any time welcoming people with no special occasion giving them a little break in everyday life.

Recall of memory : Memory Arc

Wansandong consist of old brick houses, and the senior citizen center which had been in the site was also a red brick building. Memories of the place is fragmented and stand as a small wall in every floor using red bricks.

Location 1089-16, Wansan-dong, Yeongcheon-si, Gyeongsangbuk-do, Republic of Korea **Use** Neighborhood living facilities **Site area** 115.52m² **Built area** 69.15m² **Gross Floor area** 197.79m² **Completion** 2022 **Project manager** Sangkyong Jeong, Inkeun Ryoo, Doran Kim **Structural engineer** HANGIL structural engineering **Mechanical engineer** GM EMC **Telecommunication equipment** GM engineering **Contractor** Seyoon Industrial Development Co., Ltd. **Photographer** Inkeun Ryoo

주민소통 공간 도시의 기억 : 낮은 문턱의 공공건축_공공의 뜰

〈완산뜨락 주민소통방〉은 마을의 골목 모퉁이를 환히 밝히며, 만남의 공간이자 쉼터로 이웃의 삶을 나누는 공간을 그리며 설계되었다. 이웃 간의 만남의 빈도를 생성하는 다양한 건축적 장치를 고민하였다. 작은 공공건축물로서 외부에서 더 친근하고 쉽게 접근하고, 내부의 공간이 충분히 관찰되는 것이 건물의 주요 개념이 되었다. 건축공간을 확장하고 골목길에 아이덴티티를 부여하는 골목길 디자인을 함께 제안하였다. 골목길 디자인을 통해 건축과 마을이 친밀히 엮이며 생동감 있는 흐름을 생성하는 것을 목표로 삼았다.

건축물의 도입부와 색채를 통일하여 마을의 길과 건축물이 자연스럽게 연결되도록 하였다. 골목길에 적용된 선형 패턴은 낡은 골목의 이미지를 재생성하며, 공공의 안전을 지키는 횡단보도처럼, 골목을 지나는 차량과 주민들에게 잠시 멈춤을 알리는 특별한 기호가 되어 이웃과 마을을 지킨다.

확장 : 우연한 만남 & 친근한 건물

길과 맞닿은 마을의 공간으로서 누구나 친근하게 들릴 수 있는 공간이 되길 바랐다. 골목길을 걷다 잠시 앉아 쉴 수 있는 툇마루, 비가 올 때 비를 피할 수 있는 처마 등 외부에서도 친근히 건축물을 사용할 수 있도록 한 것도 이러한 의도에서였다. 특별한 목적이 없어도 들러 책을 읽거나, 공부를 하거나, 운동을 하며 이웃을 만나는 바쁜 일상 중에 잠시 쉬어갈 수 있는 마을의 휴식처가 되도록 하였다.

기억의 소환 : 메모리 아크

마을은 오래된 벽돌집들로 이루어졌으며, 기존 경로당 또한 적벽돌의 건물이었다. 완산 뜨락 마을의 모습을 담으면서도 새로운 이미지의 공간을 바랐던 주민들의 바람을 담아, 마을을 밝히는 환한 색상의 새로운 재료를 택하면서도, 각 층의 공용공간에는 마을의 벽을 닮은 적벽돌의 '메모리 아크'를 제시하였다.

→ Corner view → Color work of the road

↑ Exterior view

NORTH ELEVATION

EAST ELEVATION

← View from southeast → View from west

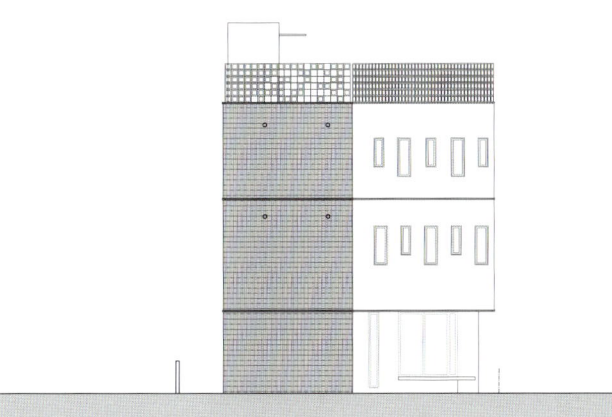

WEST ELEVATION

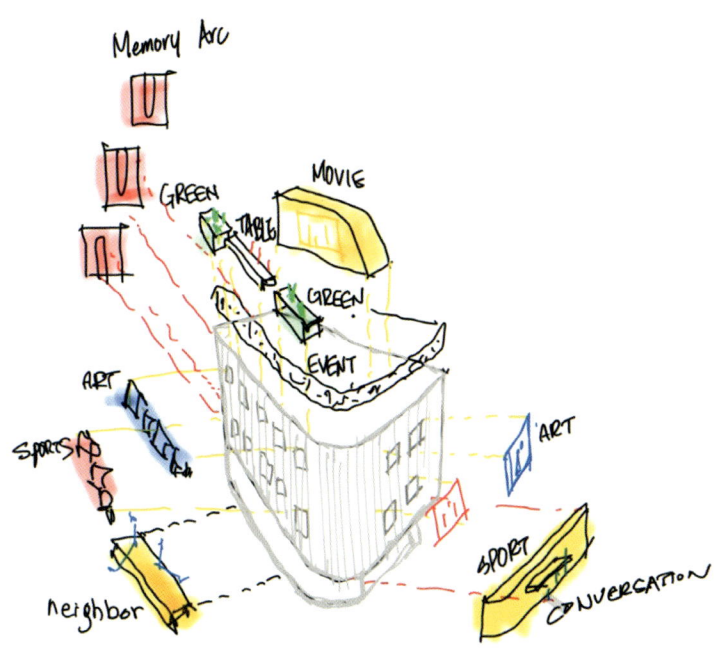

SOUTH ELEVATION

ELEMENTS SKETCH

← Piloti as a street shelter　→ Narrow wooden bench

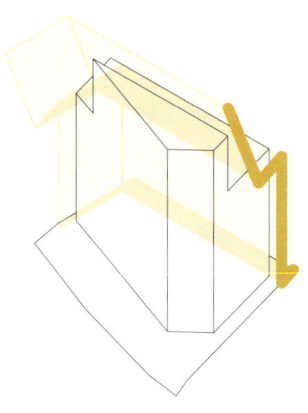

01 건축가능영역 : Regulation

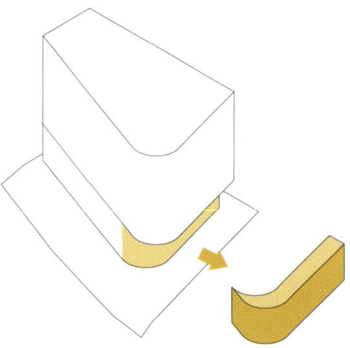

02 필로티 : Street shelter

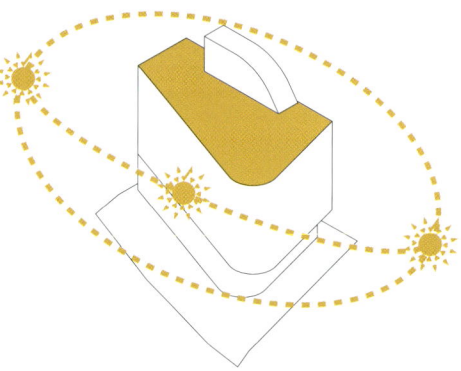

03 옥상 : Event & Green

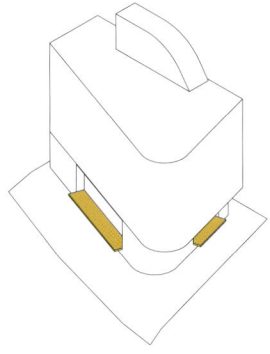

04 툇마루 : Invite

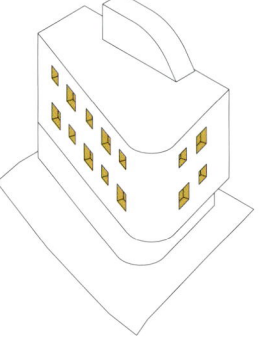

05 창 : Communication

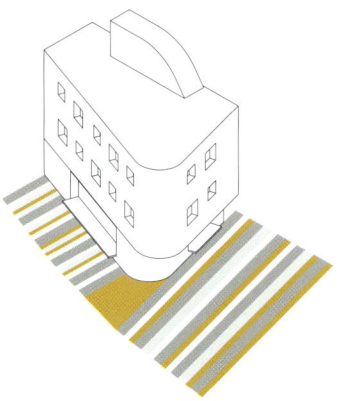

06 공간의 확장 : Street

DIAGRAM

↑ Interior view on the 1st floor   ↳ Shared kitchen

← Book cafe   → Interior view

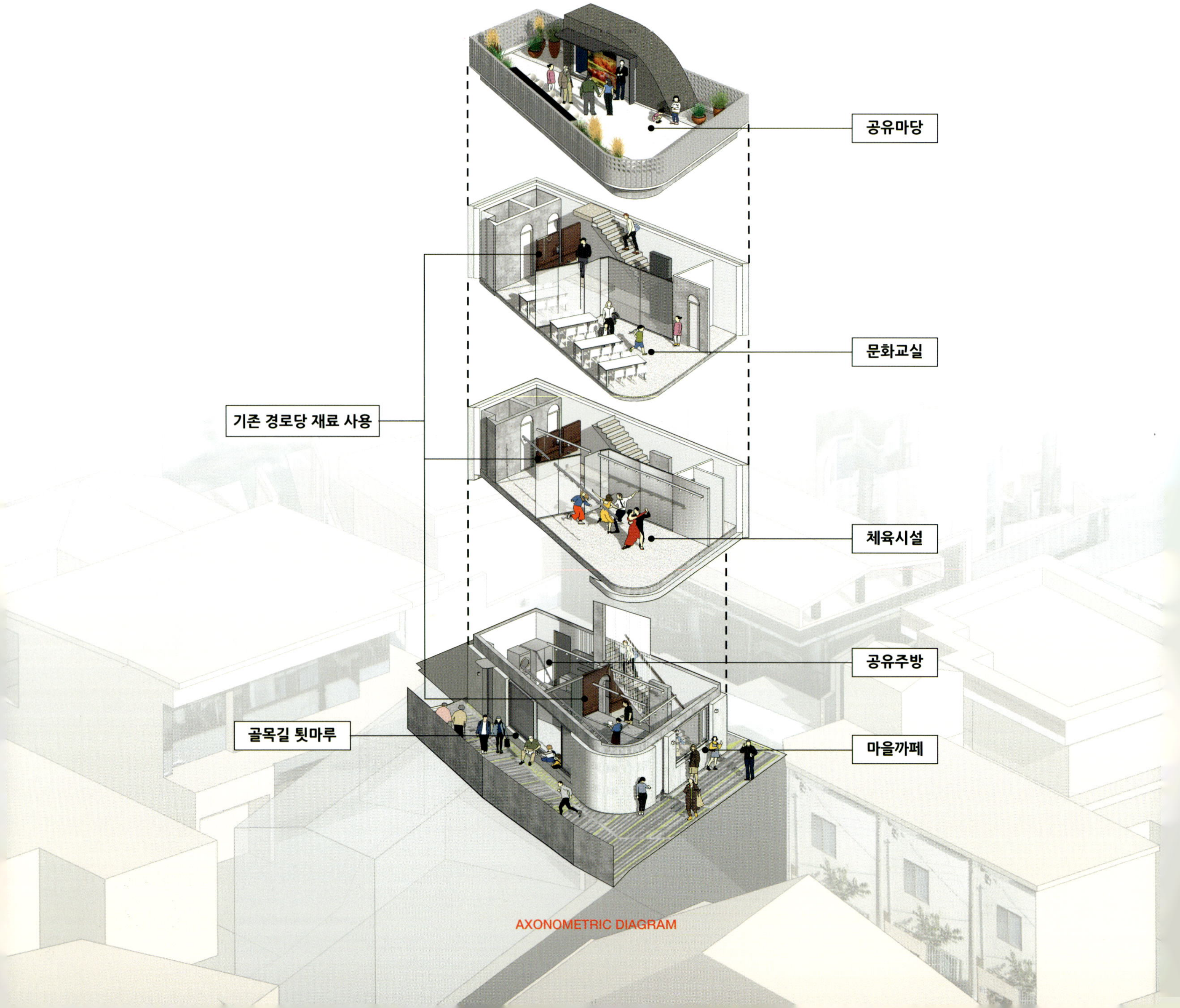

- 공유마당
- 문화교실
- 기존 경로당 재료 사용
- 체육시설
- 공유주방
- 골목길 툇마루
- 마을까페

AXONOMETRIC DIAGRAM

↑ Shared kitchen   ← Recycled brick   → Staircase

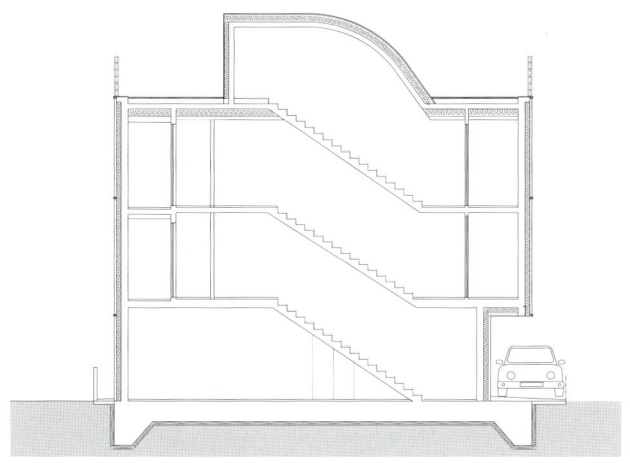

LONGITUDINAL SECTION

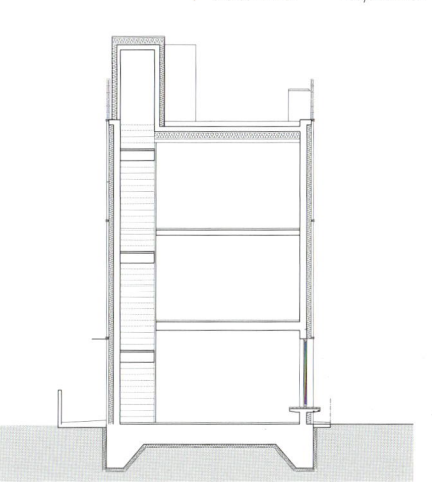

CROSS SECTION

↑ Sports space for residents   ← Meeting room for residents   ↙ Windows   → Staircase

↑ Rooftop garden

1 BOOK CAFE	4 STAIRCASE	7 CHANGING ROOM	10 STORAGE
2 SHARED KITCHEN	5 MEETING ROOM FOR RESIDENTS	8 RESTROOM	11 ROOFTOP GARDEN
3 PARKING LOT	6 SPORTS SPACE FOR RESIDENTS	9 CULTURE CLASSROOM	

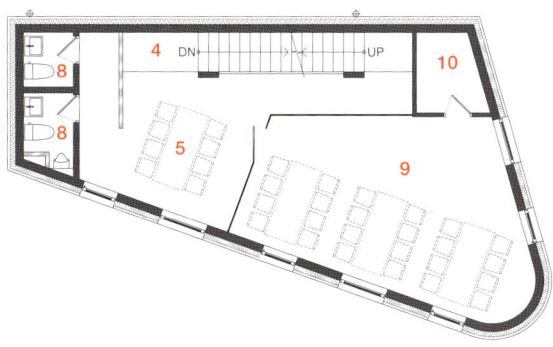

3RD FLOOR PLAN

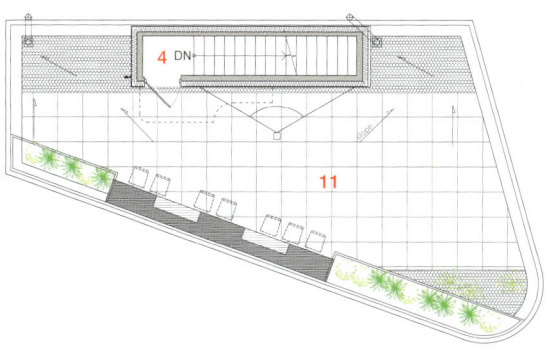

ROOFTOP FLOOR PLAN

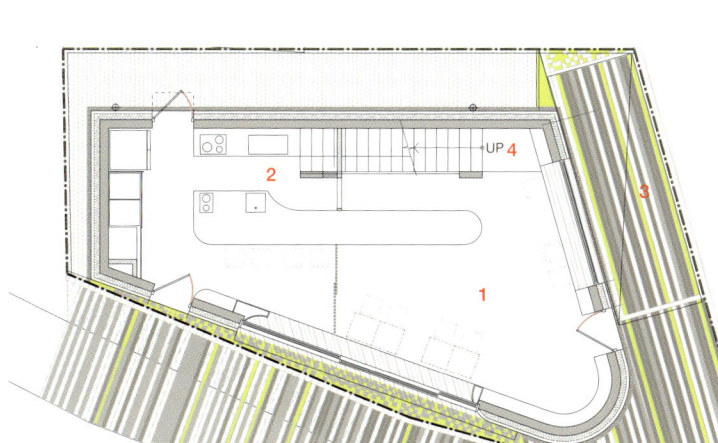

1ST FLOOR PLAN

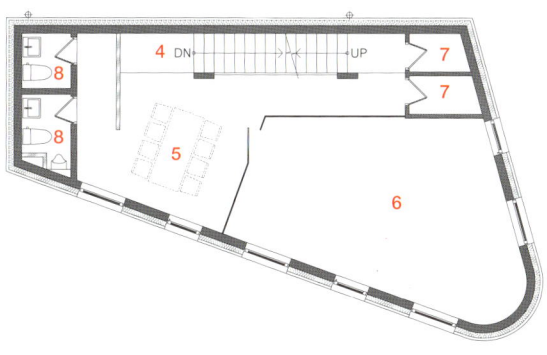

2ND FLOOR PLAN

플라잉 블록 호텔
THE FLYING BLOCK HOTEL

ARCHITECT : TAA DESIGN / NGUYEN VAN THIEN

THE CONTEXT OF THE PROJECT

The project is a complex of accommodations and resorts for professionals living and working in Phu My town, where the largest concentration of industrial parks in Ba Ria Vung Tau province is located, with many heavy industrial factories and container ports. Improving the quality of the living environment is the focus of the project.

Flying Greenery Blocks

The entire project comprises 23 apartments, featuring a diverse range of sizes from 22 to 40 square meters, including balconies and planters spanning approximately 8-12 square meters. This creates a visual effect of "the flying block" when viewed from the outside. The concept behind creating "flying greenery blocks" is to provide a variety of outdoor activities, such as a garden, elevated playground, etc., that are integrated into each space to offer a natural experience and relaxation after a day of work in the industrial park. The blocks are arranged in a staggered manner to create vertical spaces for trees to grow and develop, while also creating a lively rhythm on the facade.

Ventilation & Sunshade

The layout is divided into 4 blocks, with the traffic lane also serving as a natural ventilation gap to increase airflow. The combination of the extended blocks and large trees provide shade to the surface of the building. "The flying block" is a design solution that aims to improve the living quality of high-rise complexes in urban areas.

프로젝트의 배경

이 프로젝트는 바리어붕따우 지방의 최대 산업단지가 있는 푸미 타운에서 생활하고 일하는 전문가들을 위한 숙박 및 리조트 단지이다. 이 지역에는 다수의 중공업 공장과 컨테이너 항만이 자리하며, 프로젝트의 초점은 생활 환경의 질을 높이는 것에 맞춰져 있다.

플라잉 그리너리 블록

전체 프로젝트는 22~40㎡의 다양한 크기의 23개 객실로 구성되어 있으며, 약 8~12㎡의 발코니와 식재 공간이 포함되어 있다. 이는 외부에서 볼 때 '플라잉 블록'의 시각적 효과를 만든다. '플라잉 그리너리 블록'을 만드는 콘셉트는 각 공간에 통합된 정원과 고층 놀이터 등 다양한 야외 활동을 제공하여 산업단지에서의 하루 일과 후 자연스러운 경험과 휴식을 제공하는 것이다. 블록들은 수직 공간에서 나무가 자라고 발달할 수 있도록 교차 배치되어 있으며, 파사드에 생동감 넘치는 리듬을 연출한다.

환기 및 채광

레이아웃은 4개의 블록으로 나뉘며, 차로는 공기 흐름을 증가시키는 자연 환기의 간극 역할을 겸한다. 확장된 블록과 높다란 나무들은 한데 어우러져 건물 표면에 그늘을 제공한다. '플라잉 블록'은 도심 초고층 단지의 삶의 질 개선을 목표로 하는 디자인 솔루션이다.

Location Phu My, Ba Ria Vung Tau, Vietnam **Use** Hotel & Apartment **Site area** 300㎡ **Building area** 200㎡ **Gross floor area** 1,200㎡ **Completion** 2023 **Project manager** Nguyen Van Thien **Design team** Tran Anh Huy, Ngo Thi Bao Nhi **Contractor** Doricons **Photographer** Hoang Le

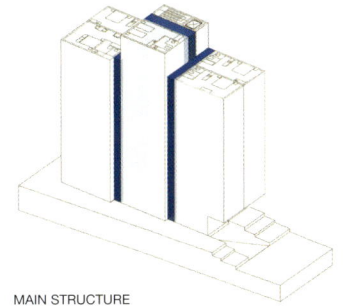

MAIN STRUCTURE

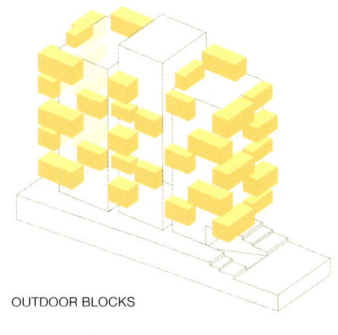

OUTDOOR BLOCKS

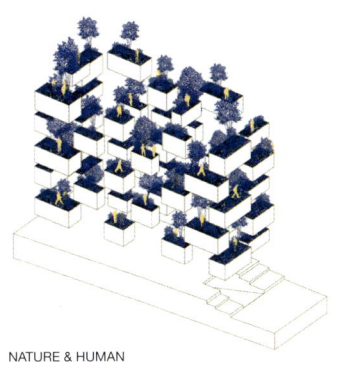

NATURE & HUMAN

DIAGRAM

← Front facade

↑ Panoramic view

IDEA DIAGRAM

↑ Exterior view

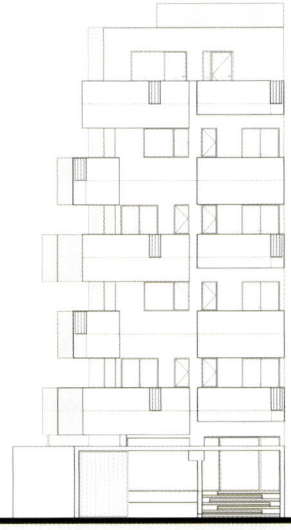

FRONT ELEVATION

SIDE ELEVATION

← Entrance   → Flying block

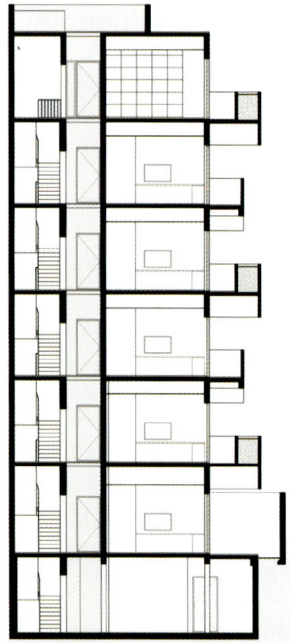 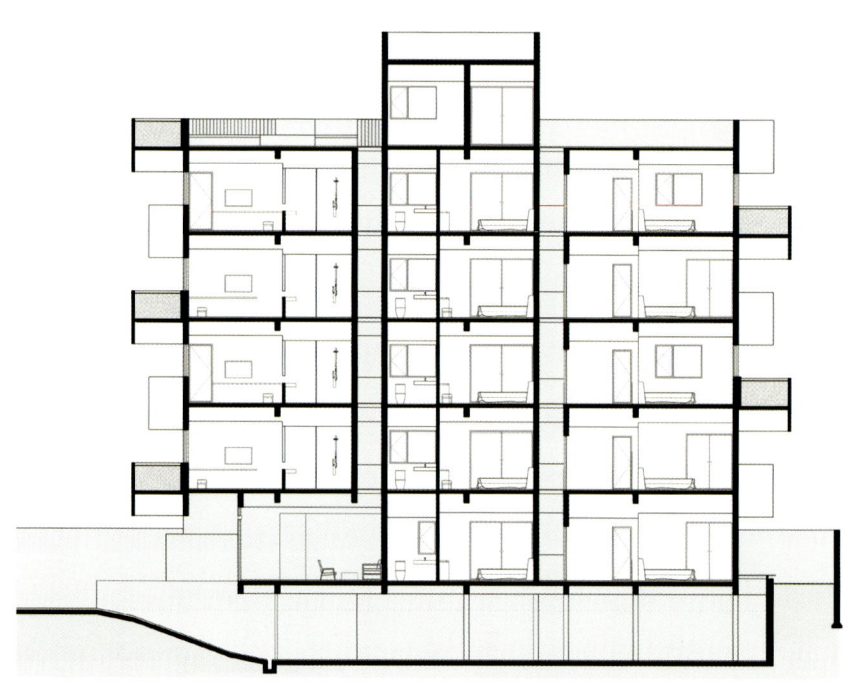

CROSS SECTION **LONGITUDINAL SECTION**

↑ Lobby ← Exterior view ↙ Exterior view → Patio

↑ Hotel room

1 ENTRANCE	4 HOTEL ROOM	7 PRIVATE GARDEN	10 LAUNDRY ROOM
2 PUBLIC GARDEN	5 CORRIDOR	8 BALCONY	11 MUTIPURPOSE ROOM
3 LOBBY & RECEPTION	6 STAIRCASE	9 TERRACE	12 WAY TO BASEMENT

2ND & 4TH FLOOR PLAN

1ST FLOOR PLAN

↖ Rooftop garden ↙ Balcony → Exterior view

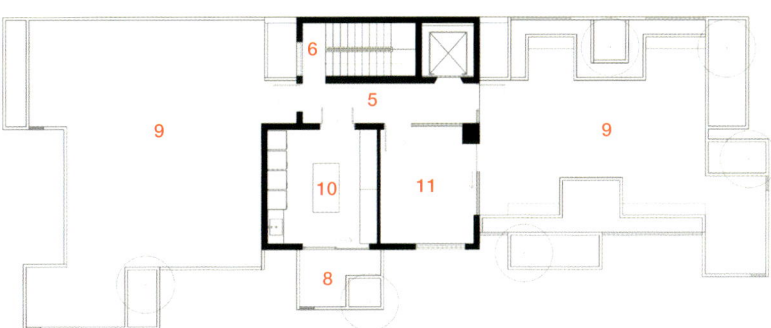

6TH FLOOR PLAN

3RD & 5TH FLOOR PLAN

161

콴 스피어
QUAN SPHERE

ARCHITECT : ÜROBROUS_STUDIOLAB / HAO-CHUN HUNG

QUAN IS LOCATED IN SANCHONGDINGKAN INDUSTRIAL ZONE, TAIWAN. It is a typical street house commonly seen in Taiwan. The building skin has experienced traces of use over time. The predecessor of QUAN was the processing and manufacturing factory of elevator axle parts. In the design and planning, the space on the third floor is set as exhibitions, lectures, and exhibitions, and the elevator the industry combines art, design, manufacturing and other multi-functional space performances to expand different customer groups and industrial types. The second floor is a reception, office and meeting space, and the first floor is a new type of storage and shipping space.

Under the discussion between the design team and the owner, the concept of "new type factory" was taken as the deconstruction and reintegration of ordinary materials, and integrated into the contemporary design and the surrounding environment. The mushroom head, outdoor unit, and water tower on the iron roof, stated Out of the local characteristics, we hung a large area of orchid mesh cloth, which is common in agriculture, on the facade of the building, reflecting the surrounding environment like a canvas, and filtering those objects that had to be covered in the past, making it a framed scene, hoping to not distract the guest. The interior and exterior of the building are slightly isolated with thin and transparent materials, thereby creating two viewing angles.

- From inside to outside : The mesh cloth on the exterior facade solves the privacy problem of the large floor-to-ceiling windows. What's more interesting is that standing in front of the large window, the mesh cloth seems to have a layer of filters for the urban street scene, and the wind and the rain will be a bunch. Light can become a painting.
- From outside to inside : The silver-gray orchid mesh with metallic luster is chosen for the design. A pure and elegant face that reflects the surrounding environment like a canvas. The mesh cloth on the long facade also turns the original facade of QUAN into a silhouette, which is vaguely hidden behind the canvas. It conforms to the surrounding urban environment. Temperament, the smell and sound of the city, plus the blowing of the wind and the floating of the mesh, are always springs. The field brings out a different look.

On the short facade, the iron structure is used to match the opening of the facade to make cuts. The vertical and horizontal line segmentation echoes the telephone poles and wires on the street on the side, which strengthens the surrounding environment. The intricate, vigorous and casual urban state. The design method of the short facade window frame, using metal plate to draw the window frame to adjust the light in the west. Also extends the visual experience of the window scene, invites the surrounding environment to enter the space, and re-chews the original flavor belonging to Taiwanese industrial area. The experimental elevator that runs through from ground to top floors where is located in the center of the base and is enclosed in a transparent glass curtain. The glass reflects the environment, exposes the elevator parts, and the vertical process of people traveling up and down in the light elevator, it achieves the effect of displaying products/space at the same time.

Location New Taipei City, Taiwan **Use** Factory, Office **Built area** 176.6m² **Gross area** 530m² **Completion** 2022 **Project manager** Hao-Chun Hung **Design team** Yaun-Yun Huang, Hong-Wei Huang, Chi-Yu Lai, Szu-Ying Tan, Shao-Bo Wu **Structure** COLUMBUS STRUCTURAL **Contractor** WEN-HUA CONSTRUCTION **Photographer** Yi-Hsien Lee and Associates YHLAA

← Exterior & shipping area   → Exterior night view

↑ South exterior view & shipping area
↓ South exterior & street view

NORTH ELEVATION

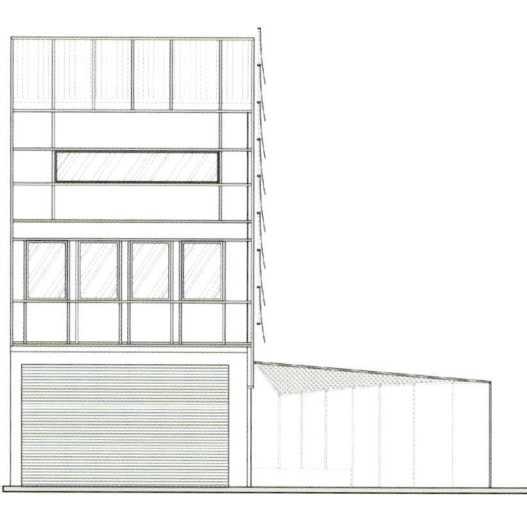

SOUTH ELEVATION

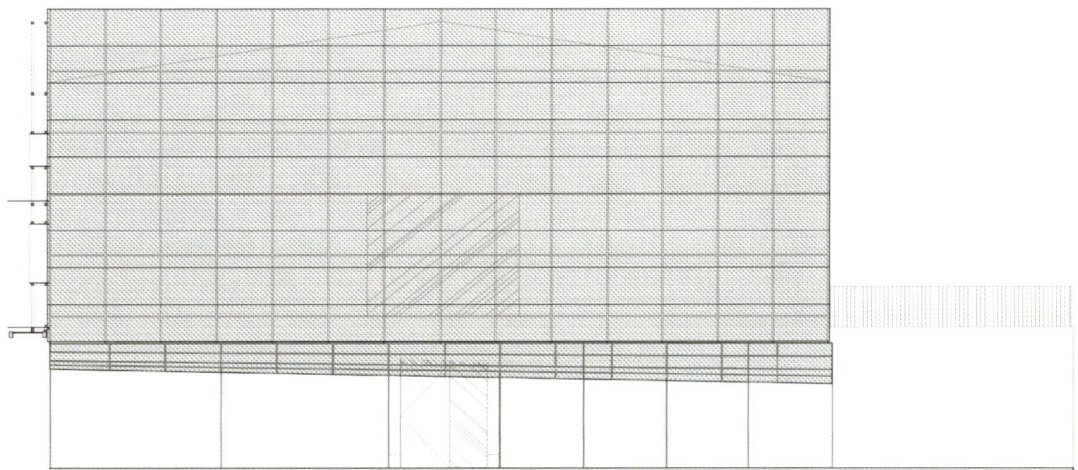

EAST ELEVATION

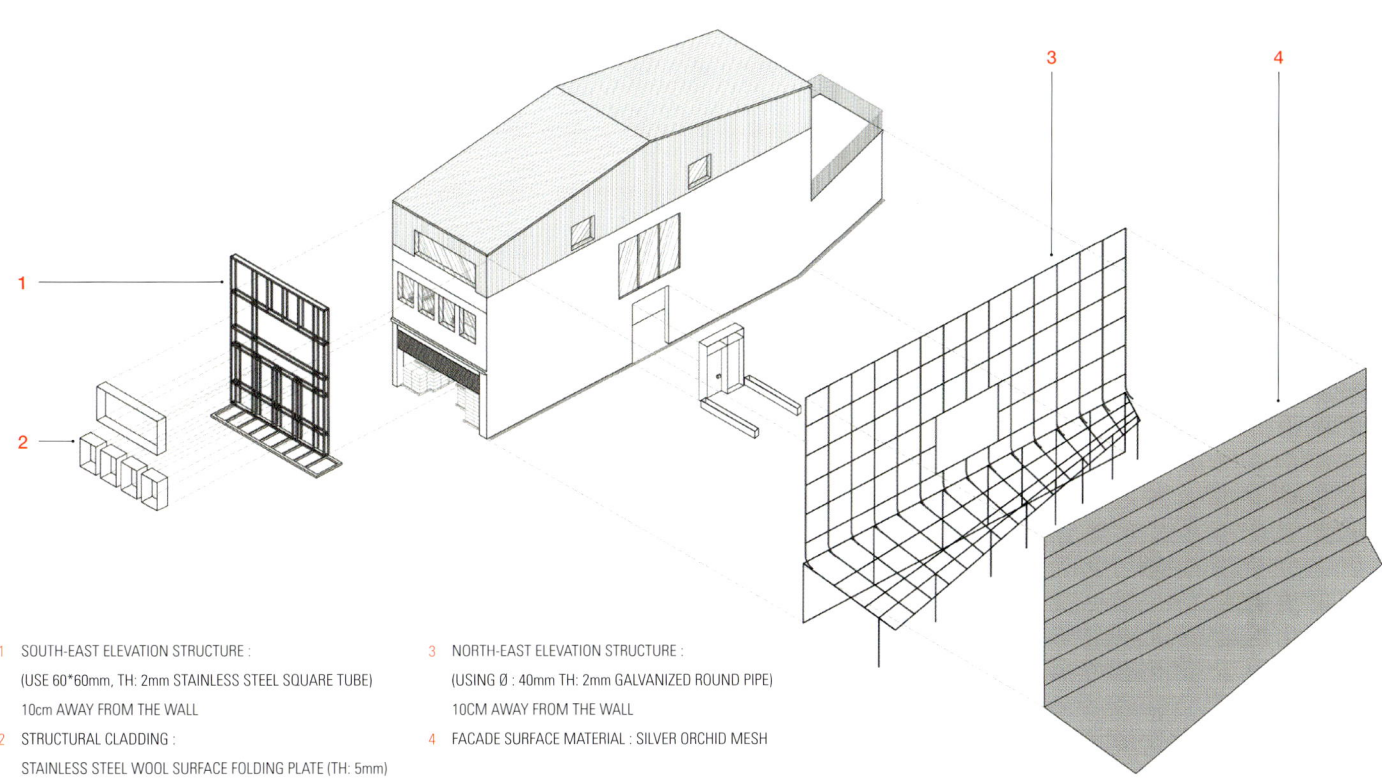

1. SOUTH-EAST ELEVATION STRUCTURE :
   (USE 60*60mm, TH: 2mm STAINLESS STEEL SQUARE TUBE)
   10cm AWAY FROM THE WALL
2. STRUCTURAL CLADDING :
   STAINLESS STEEL WOOL SURFACE FOLDING PLATE (TH: 5mm)
3. NORTH-EAST ELEVATION STRUCTURE :
   (USING Ø : 40mm TH: 2mm GALVANIZED ROUND PIPE)
   10CM AWAY FROM THE WALL
4. FACADE SURFACE MATERIAL : SILVER ORCHID MESH

FACADE PERSPECTIVE

콴은 대만 산총딩간 산업 구역에 있다. 대만에서 흔히 볼 수 있는 전형적인 거리의 주택이다. 건물 외피는 시간이 지남에 따라 사용의 흔적이 있다. 콴의 전신은 엘리베이터 축 부품의 가공 및 제조 공장이었다. 디자인과 계획에서 세 번째 층은 전시회, 강연, 그리고 전시를 위한 공간으로 설정되며, 엘리베이터 산업은 예술, 디자인, 제조 등 다양한 기능을 결합한 공간 연출로 서로 다른 고객 그룹과 산업 유형의 확장을 목표로 하고 있다. 2층은 접수, 사무실 및 회의 공간으로, 1층은 새로운 유형의 저장 및 배송 공간으로 구성되었다.

디자인 팀과 소유주 간의 논의 하에, "새로운 형태의 공장"이라는 개념은 일반적인 재료의 해체와 재통합으로 채택되었으며, 현대적 디자인과 주변 환경에 통합되었다. 철제 지붕 위에 버섯형 머리, 외부 장치, 그리고 물탑은 지역의 특색을 나타낸다. 건축가는 농업에서 흔히 사용되는 난초 망사 천을 건물의 파사드에 대규모로 매달아, 주변 환경을 캔버스처럼 반영하고 과거에 반드시 가려야 했던 사물을 필터링하여 액자 속 풍경을 만들어 내고자 했다. 이는 고객의 주의를 산만하게 하지 않기를 바라는 마음이었다. 건물의 내외부는 얇고 투명한 재료로 살짝 분리되어 그 결과 두 가지 시점을 생성한다.

- 내부에서 외부로: 외부 파사드에 설치된 망사 천은 바닥부터 천장까지 이어진 큰 창문의 프라이버시 문제를 해결했다. 더욱 흥미로운 점은, 큰 창문 앞에 서 있을 때, 망사 천이 도시 거리 풍경에 대한 필터의 역할을 하는 것처럼 보인다는 것이다. 바람과 비가 모여 빛이 되어 한 폭의 그림이 될 수 있다.

- 외부에서 내부로: 디자인을 위해 금속 광택이 있는 은회색 난초 망사를 선택했다. 캔버스처럼 주변 환경을 반영하는 순수하고 우아한 외관을 가진다. 긴 파사드에 걸린 망사 천은 콴의 원래 외관을 실루엣으로 변화시키며, 그것은 캔버스 뒤에 희미하게 숨어 있다. 이는 주변 도시 환경과 조화를 이룬다. 도시의 냄새와 소리, 바람의 부는 것과 그물망이 흔들리는 것은 항상 봄을 연상시킨다. 이 필드는 늘 다른 모습을 드러낸다.

짧은 파사드에서는 철제 구조물을 활용하여 파사드의 개구부와 일치시켜 절단한다. 수직과 수평 라인 분할은 거리의 전신주와 전선을 반영하여 주변 환경을 강화한다. 그것은 복잡하고 활력 넘치며 격식 없는 도시 상태를 나타낸다. 짧은 파사드 창틀의 디자인 방법은 금속 판을 사용하여 창틀을 그리고 서쪽의 빛을 조절한다. 또한 창문 풍경의 시각적 경험을 연장해 주변 환경을 공간 안으로 초대하고, 대만 산업 지역에 속한 원래의 풍미를 다시 씹어본다. 실험적인 엘리베이터는 바닥부터 최상층까지 관통하며, 기반의 중심에 있고 투명한 유리 커튼으로 둘러싸여 있다. 유리는 주변 환경을 반영하고, 엘리베이터의 부분노출과 가벼운 엘리베이터 내에서 사람들이 오르내리는 수직적 과정을 보여준다. 이는 제품과 공간을 동시에 전시하는 효과를 얻을 수 있다.

↑ North entrance & parking space

1 SHIPPING AREA	4 ELEVATIOR	7 MEETING AREA	10 MULTIFUNCTIONAL ACTIVITY AREA
2 ENTRANCE HALL	5 OFFICE	8 OUTDOOR TERRACE	
3 STORAGE AREA	6 RECEPTION	9 ARTISTRY PERFORMACE AREA	

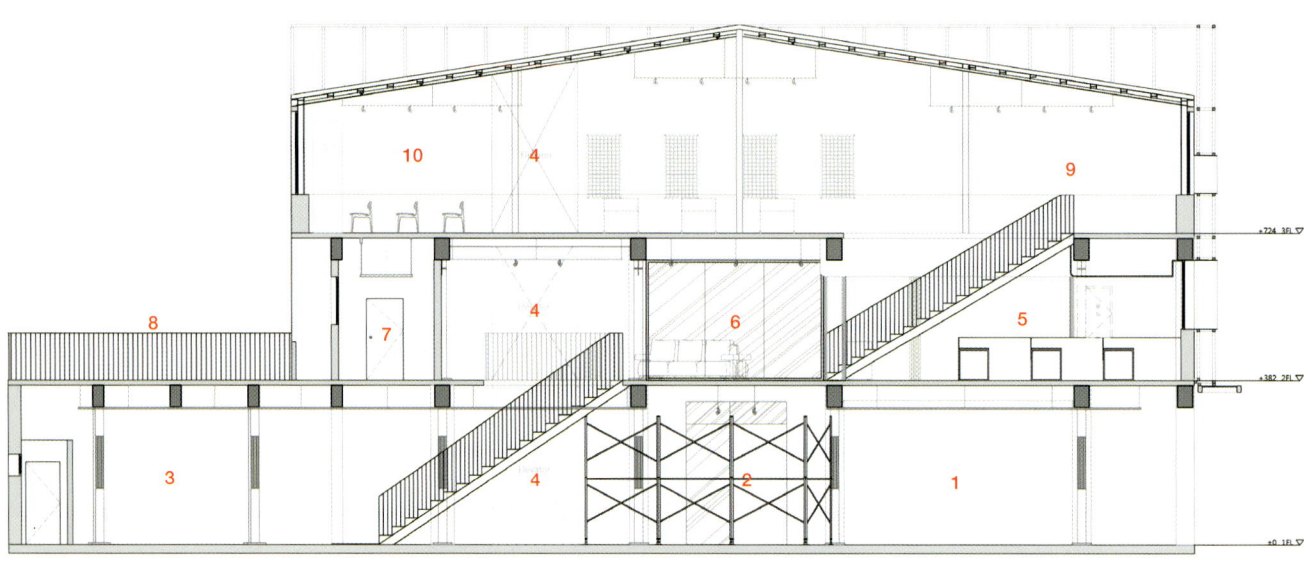

NORTH-EAST SECTION

RF PERSPECTIVE SECTION

3F Space function
Exhibition area (84m²)
Multivunctional activity area (42m²)
Elevator (5m²)

2F PERSPECTIVE SECTION

2F Space function
Office (84m²)
Reception area (42m²)
Meeting area (16m²)
Outdoor terrace (38m²)
Elevator (5m²)
Canopy (68m²)

1F PERSPECTIVE SECTION

1F Space function
Shipping area (59m²)
Entrance hall (37m²)
Storage area (44m²)
Elevator (5m²)
Parking garage (28m²)

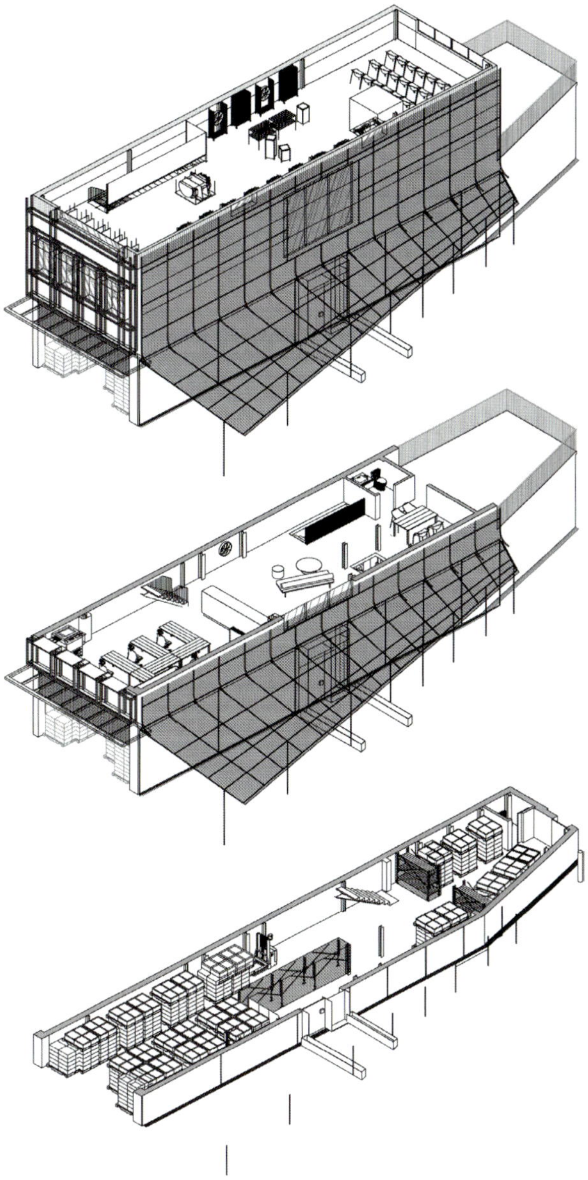

FACADE STRUCTURE EXPLOSION DIAGRAM

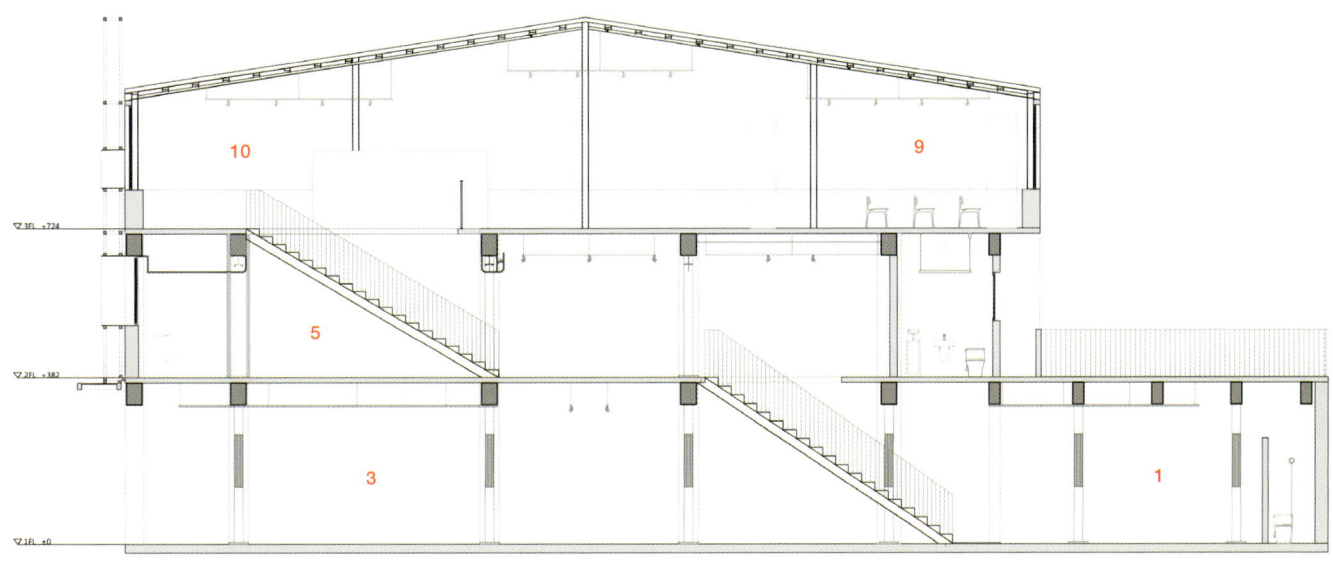

SOUTH-WEST SECTION

← Entrance & parking space night view

1 IRON FILINGS BRICK
2 TH: 3mm STAINLESS PLATE
3 TH: 3mm MIRROR STAINLESS PLATE
4 50X50mm STAINLESS SQUARE TUBE
5 MAIN DOOR HANDLE
6 HANDLE BOX CONTAINS IRON FILINGS: DIMENSION: 170X170mm
8 MAIN STRUCTURE DISTANCE: 60X60mm, TH: 2mm STAINLESS SQUARE TUBE 10cm DISTANCE FROM THE WALL
9 FULLY WELDED
10 TH: 5mm STAINLESS STEEL PLATE.
11 CLADDING IN STRUCTURE
12 TH: 20mm TEMPERED GLASS CANOPY
13 CANOPY: 100X150mm, TH 6mm STAINLESS STEEL SQUARE TUBE
14 STAINLESS STEEL PLATE
15 CANOPY: TH: 20mm TEMPERED GLASS
16 CANOPY STRUCTURE: 100X150mm, TH: 6mm STAINLESS SQUARE TUBE

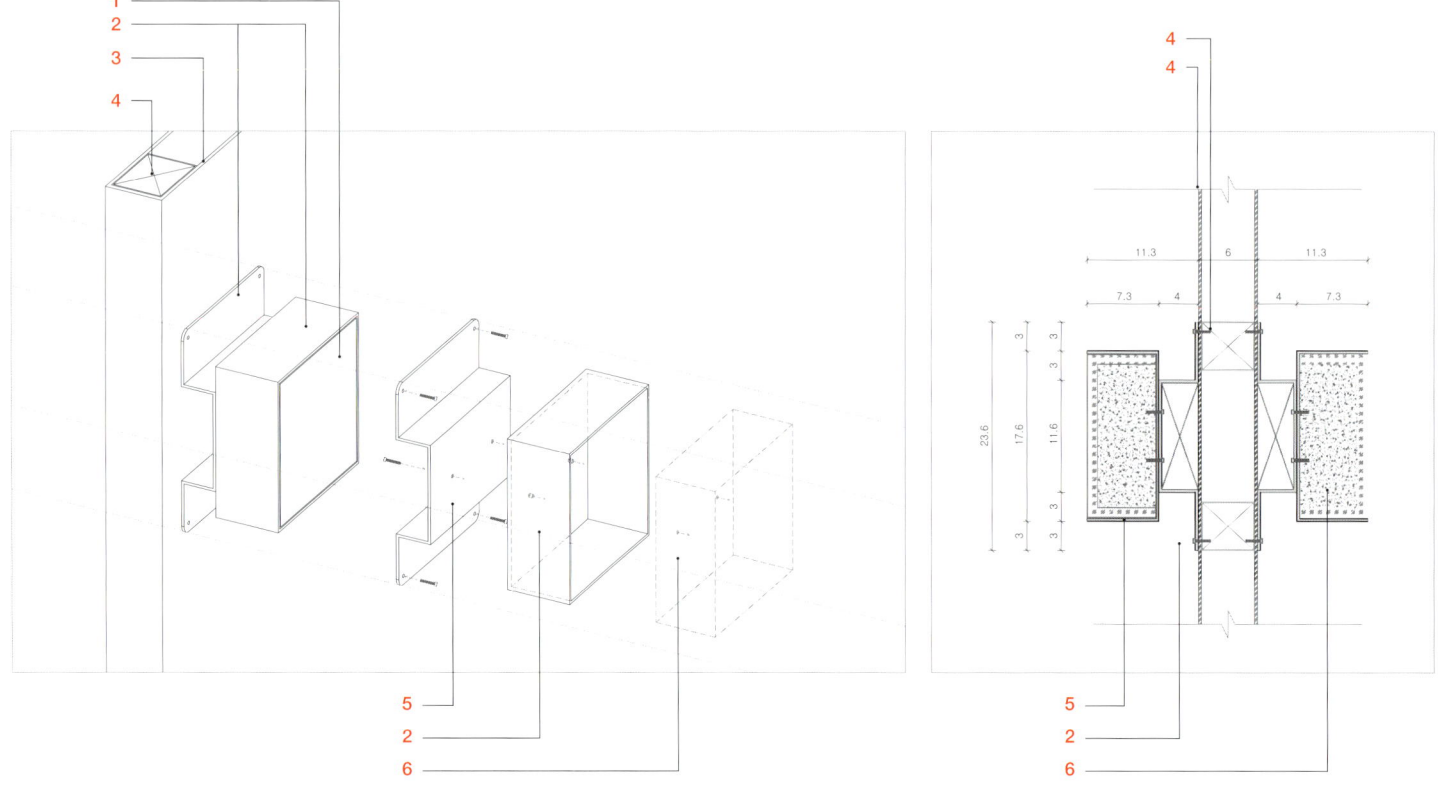

NORTH-EAST MAIN DOOR HANDLE DETAIL DRAWING

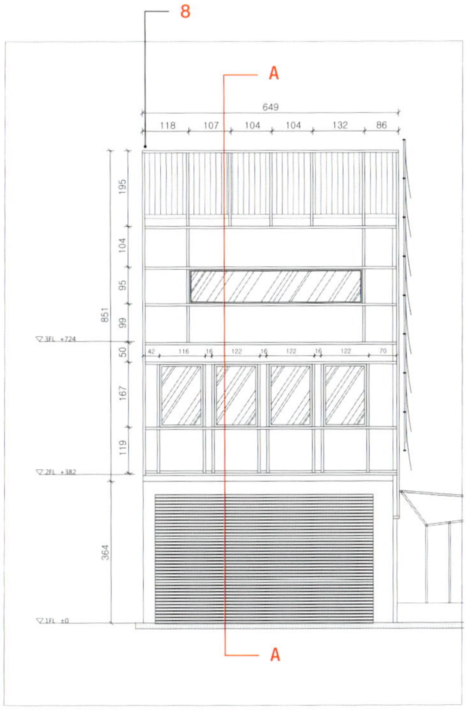

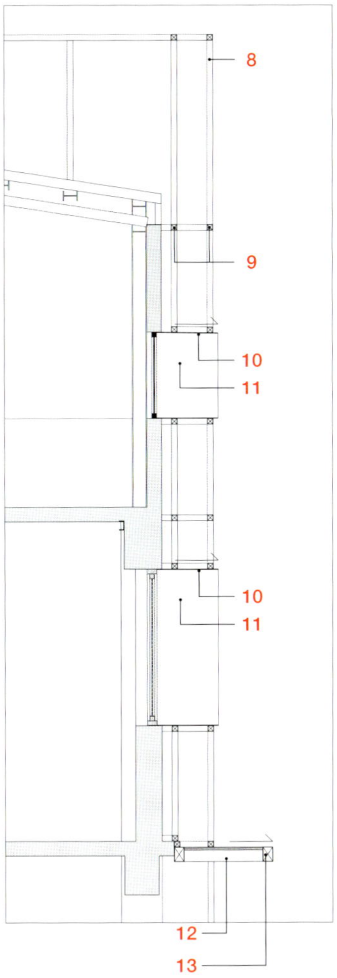

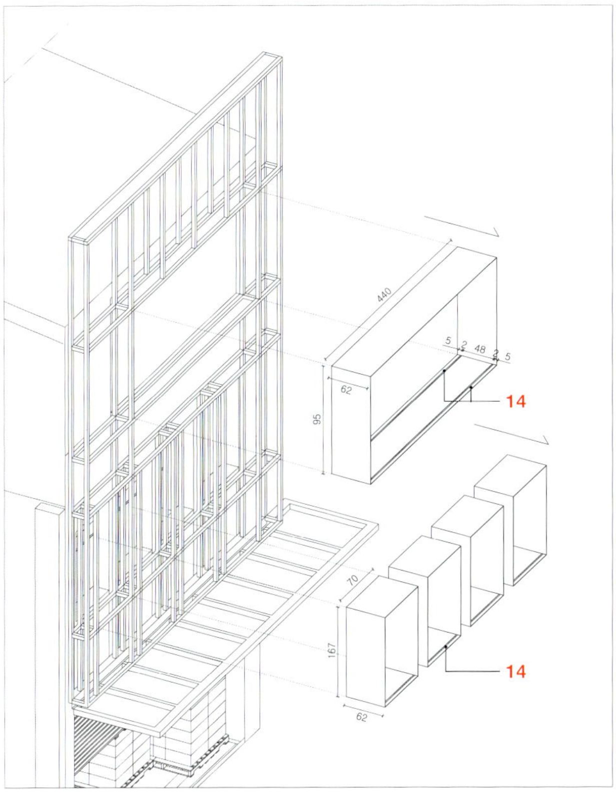

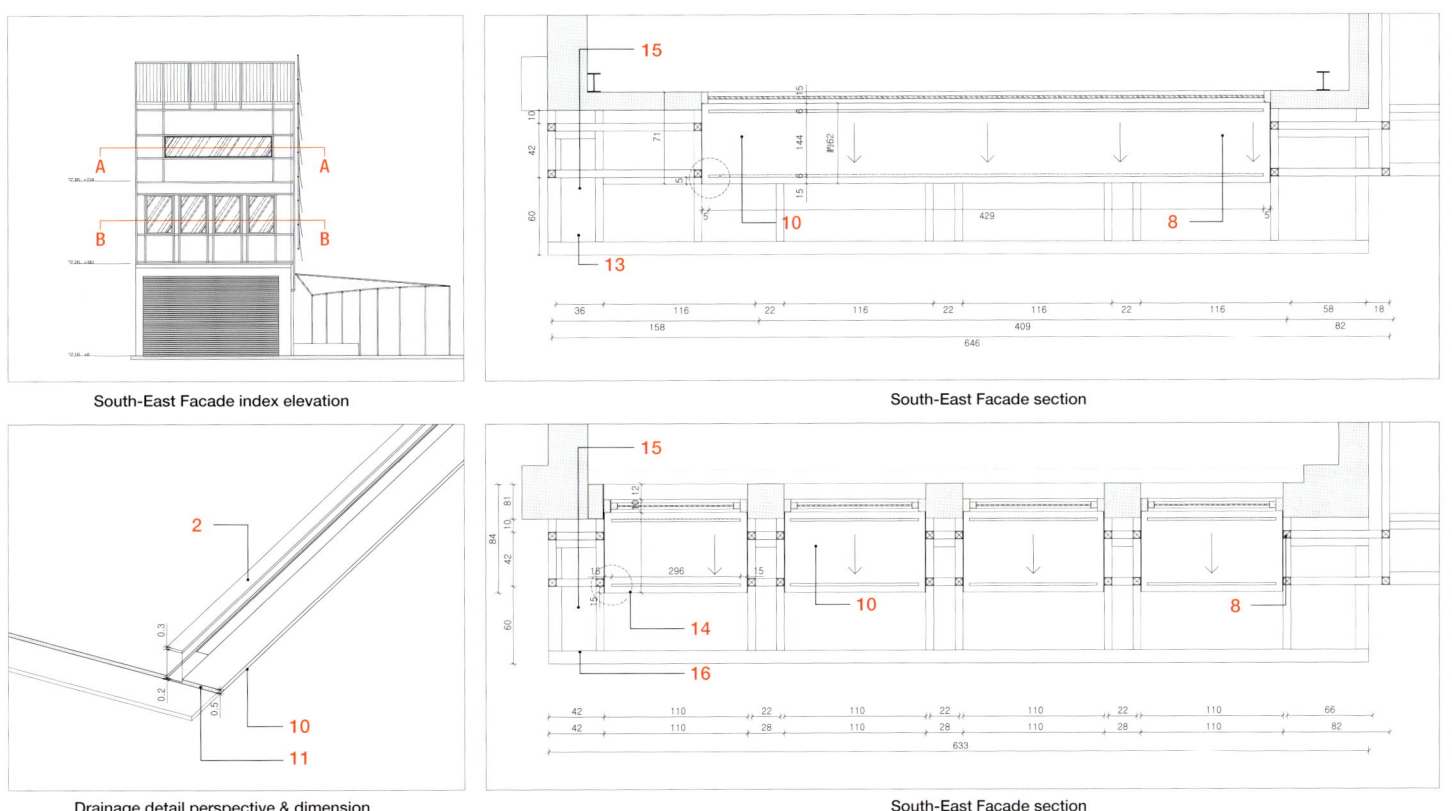

SOUTH-EAST FACADE STRUCTURE DETAIL

South-East Facade index elevation

South-East Facade section

Drainage detail perspective & dimension

South-East Facade section

SOUTH-EAST FACADE STRUCTURE DETAIL

↑ 1st floor space   ← North enterance & garden   ↗ 1st floor enterance hall   ↘ 1st floor storage

← 2nd floor & 3rd floor stair   → Stair detail

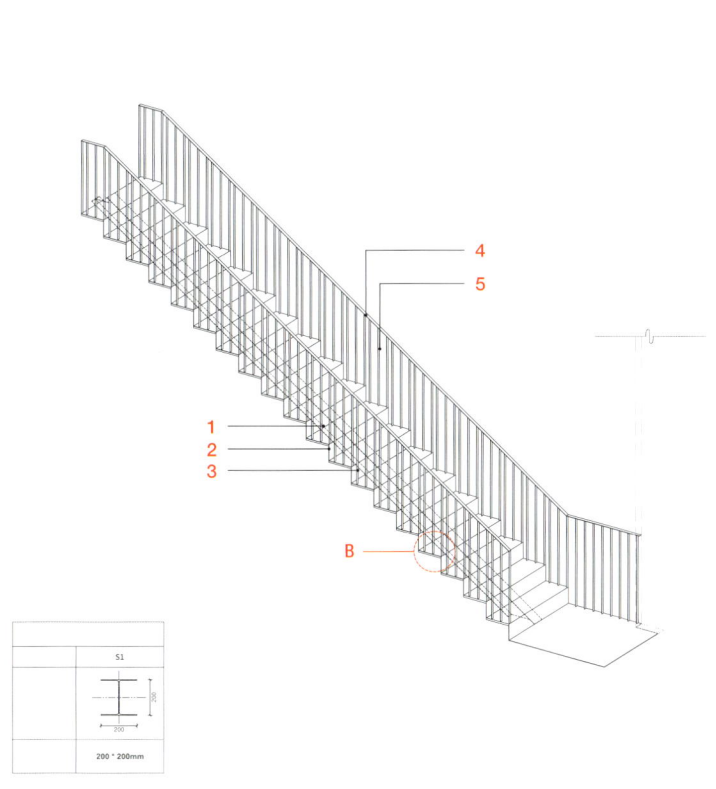
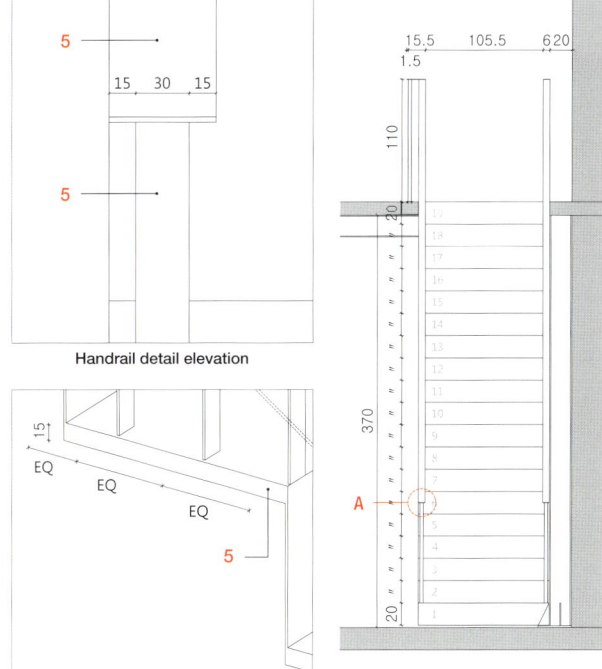

Handrail detail elevation

Handrail detail perspective

Staircase elevation

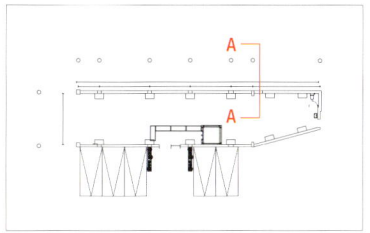

1 200X200 H BEAM STRUCTURE
2 TH: 3mm, STAINLESS PLATE FULLY
3 WELDED STAIRS
4 TH: 3mm, W: 60mm STAINLESS PLATE HANDRAIL
5 TH: 3mm, W: 30mm STAINLESS PLATE COLUMN

STAINLESS STEEL STAIRCASE DETAIL

← 2nd floor view　→ 2nd floor window view　↙ 2nd floor office view　↳ 2nd floor reception area

1	SHIPPING AREA	4	ELEVATIOR	7	MEETING AREA	10	MULTIFUNCTIONAL ACTIVITY AREA
2	ENTRANCE HALL	5	OFFICE	8	OUTDOOR TERRACE	11	TOILET
3	STORAGE AREA	6	RECEPTION	9	ARTISTRY PERFORMACE AREA	12	PARKING SPACE

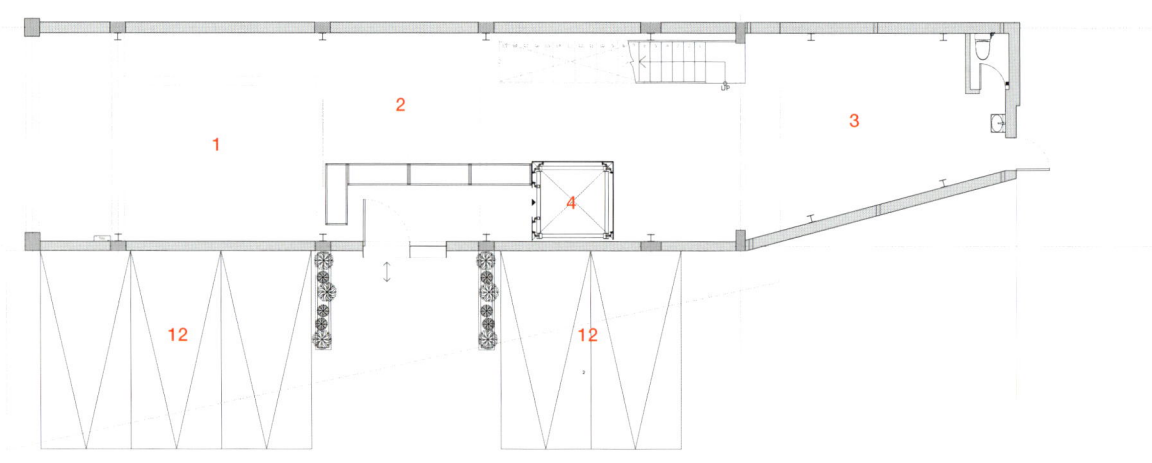

1ST FLOOR PLAN

↑ 3rd floor artistry performance area

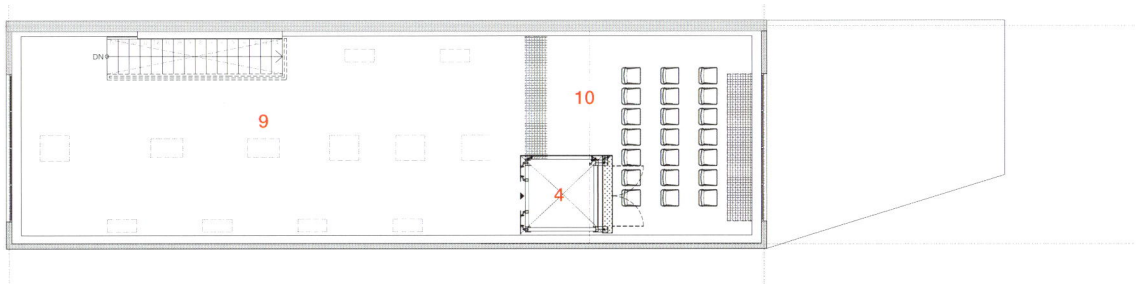

3RD FLOOR PLAN

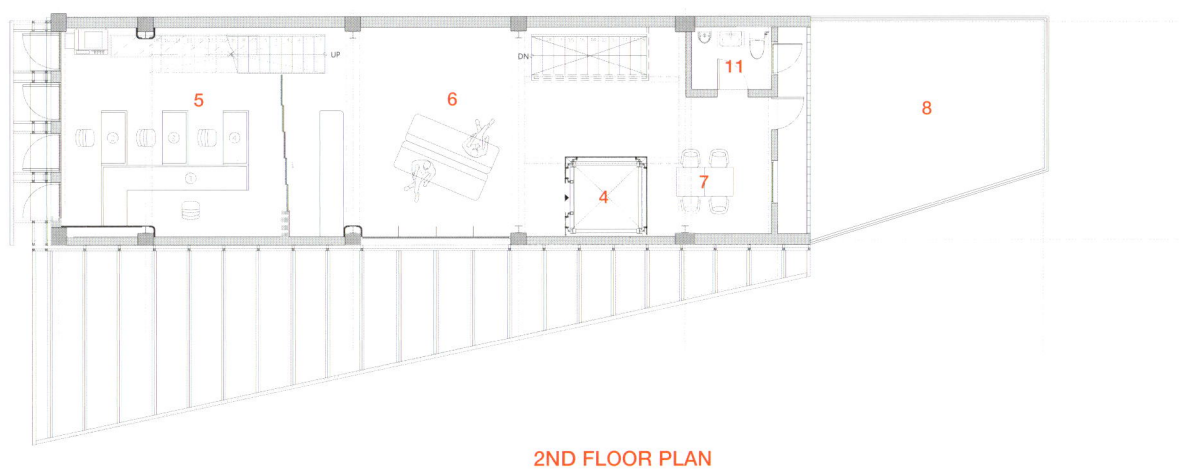

2ND FLOOR PLAN

얼라이브 레지던스
ALIVE RESIDENCE

ARCHITECT : SATA NA ARCHITECT / CHALERMCHAI ASAYOTE

COMBINING THE BALANCE OF CITY LIVING TOGETHER WITH NATURE. Home design ideas with natural light and breeze. Leaving space around the house for a garden area, including inserting plants on the balcony to create a good atmosphere for living. It is the heart of home design. The design process begins by considering the limited usable space and modification to support the space adjustment to future lifestyle changes. The living space is designed to accommodate any lifestyle. Designing the space to suit growing families involves creating areas for in-home work and responding to work-from-home or changes for the home office setup.

Turning away from the road is the primary concept. in organizing living space. This house was intentionally designed as a relaxation space for people. When opening the door, the primary desire is to disconnect from the mess of the street. It will make that like a space of peace and relaxation when coming home. The building's appearance is showcased through a three-story structure enveloped in a wooden facade and wooden slats, crafted with the concept of integrating wood. The decorative wood contains aluminum with a wood pattern and artificial wood, obstructing external views for a sense of privacy. This choice also imparts a natural ambiance and connections to traditional wooden houses. On the balcony, planted plants act as a pocket garden, filtering light and dust for a pleasant atmosphere for the residents.

Creating a separation from the road and experiencing nature with a wooden facade is essential in distinct feelings. This concept is key to making the house a residence as a starting point for life and a beginning of inspiration for urban living. It's designed to be a space that holds good memories and happiness whenever you come home.

Location Bangkok, Thailand **Use** Housing **Site area** 140m² **Building area** 237m² **Gross floor area** 237m² **Completion** 2023 **Project manager** Chalermchai Asayote **Photographer** Rungkit Charoenwat

도시 생활과 자연의 균형을 조합한 얼라이브 레지던스는 자연광과 바람을 활용한 주거 디자인 아이디어를 중심으로 한다. 주거 공간 설계의 핵심은 살기 좋은 분위기를 조성으로, 집 주변에 정원 공간을 남겨두고 발코니에 식물을 심었다. 설계 과정은 사용 가능한 공간의 제한과 미래의 생활 방식 변화에 대응할 수 있도록 공간 변형의 가능성을 내포하며 시작되었다. 거주 공간은 어떠한 생활 방식에도 적합하도록 설계되었다. 성장하는 가족에 맞게 공간을 디자인하는 것은 재택근무 또는 홈 오피스 설정의 변화에 대응하는 것 역시 포함한다.

주거 공간을 구성하는 데 있어, 도로와 거리를 두는 것은 주요 개념이다. 이 집은 의도적으로 거주자들을 위한 휴식 공간으로 설계되었다. 문을 열 때 주요한 바람은 거리의 혼란에서 벗어나는 것이었다. 이로써 집은 평화롭고 휴식의 공간이 된다. 건물의 외관은 목재 파사드와 목재 슬랫으로 감싸인 3층 구조를 통해 보여주며, 목재를 통합하는 콘셉트로 제작되었다. 장식용 목재에는 나무 패턴이 있는 알루미늄과 인조 목재가 포함하여 외부 시야를 차단하고 사생활을 보호해준다. 또한 이러한 선택은 자연스러운 분위기와 전통적인 목조 주택과의 연결성을 부여한다. 발코니에 심은 식물들은 포켓 가든의 역할을 하여 빛과 먼지를 걸러내는 등 거주자들에게 쾌적한 분위기를 제공한다. 도로로부터의 분리와 목재 파사드를 통한 자연 경험은 독특한 분위기를 자아내는 데 중요하게 작용한다. 이 콘셉트는 집을 삶의 출발점이자 도시 생활에 대한 영감의 시작으로 만드는 핵심이다. 집에 돌아올 때마다 좋은 추억과 행복을 간직하는 공간으로 설계되었다.

← Corner view   → Night view

↑ Panoramic view

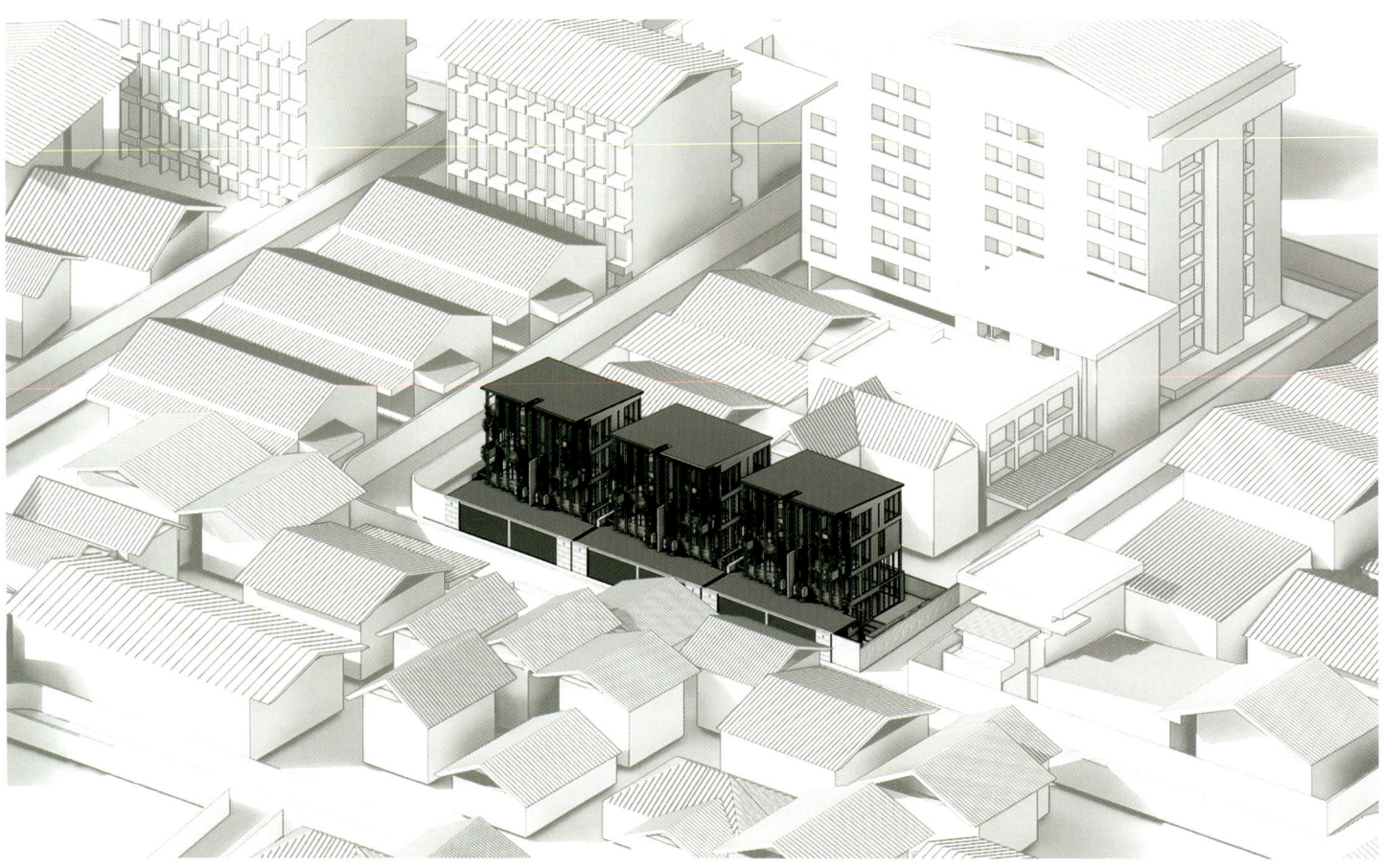

ISOMETRIC

↑ Exterior view

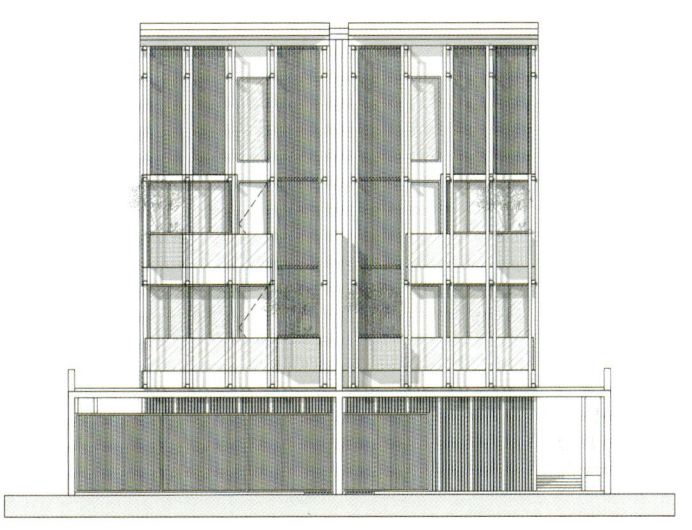

FRONT ELEVATION

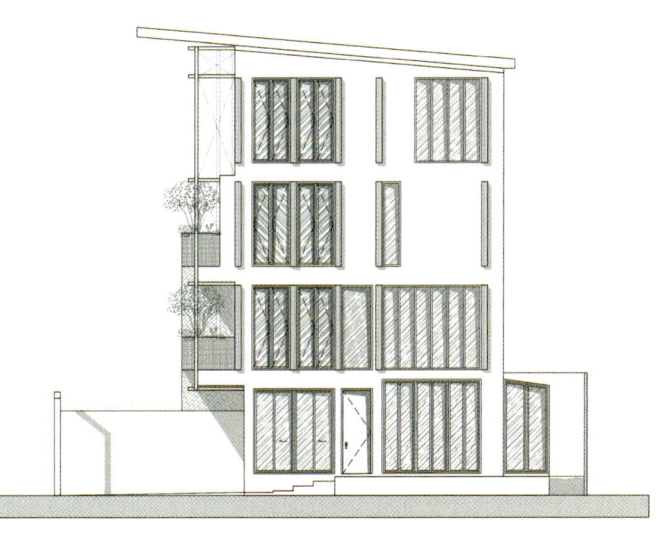

SIDE ELEVATION

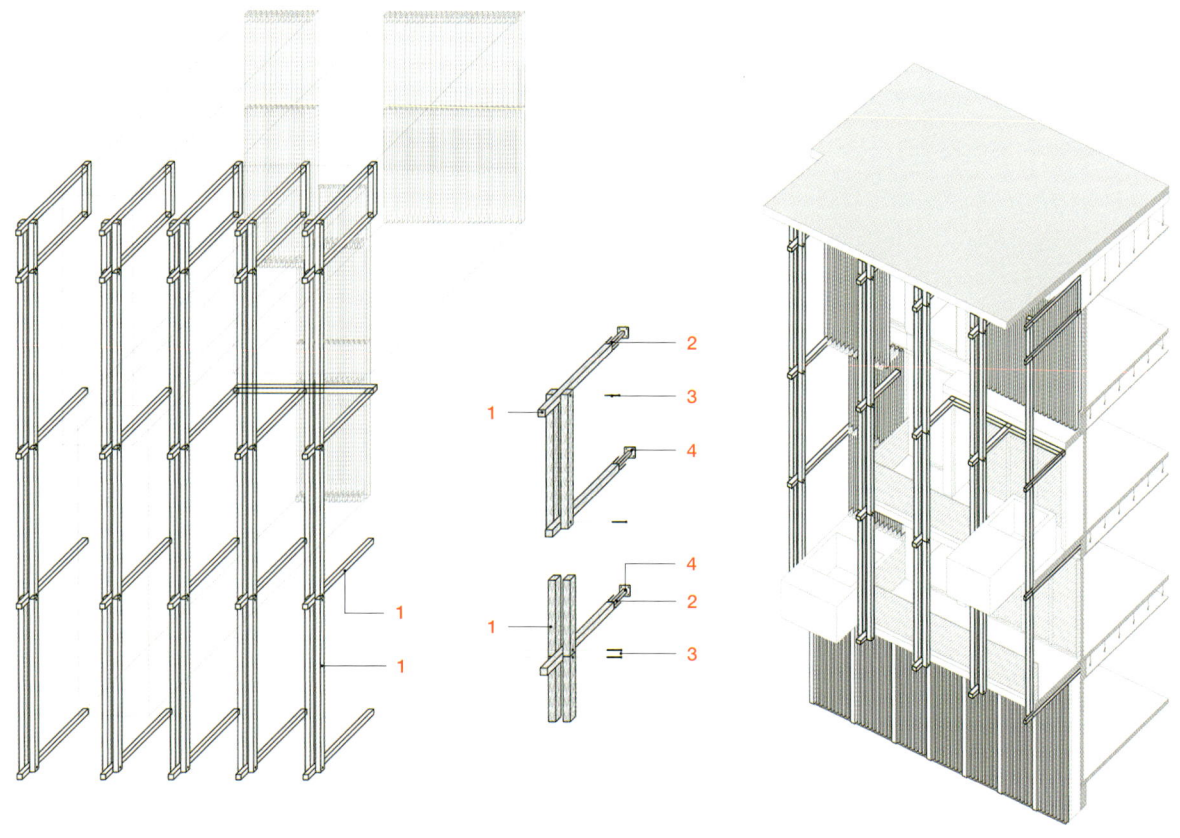

← Wooden facade & Wooden slats   → Wooden facade & Wooden slats

DETAIL OF THE FRAME

↑ Swimming pool

1	4"*4" ALUMINUM WOOD FINISH POWDER COATION	6 PANTRY	11 BEDROOM
2	STEEL SUPPORT BEAM	7 DINING ROOM	12 WALK-IN CLOSET
3	HEAVY DUTY SCREW	8 LIVING ROOM	13 BATHROOM
4	SQUARE STEEL PLATE	9 TERRACE	14 STUDY
5	PARKING LOT	10 BAR	

SECTION

Interior view
Living room
Sofa area

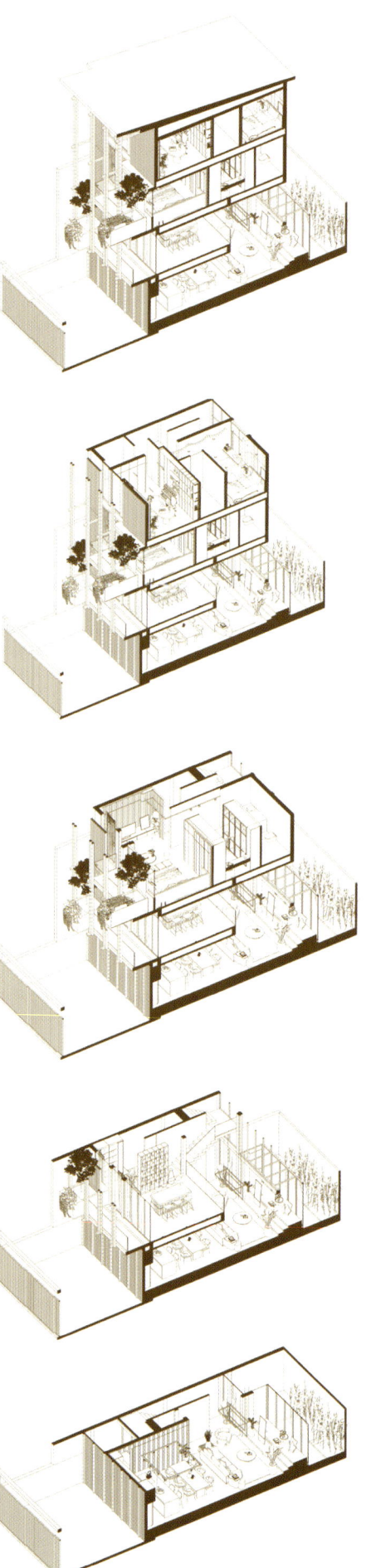

ISOMETIRC

← Home bar → Dressing table

1	PARKING LOT	4 DINING ROOM	7 TERRACE	10 WALK-IN CLOSET
2	LAUNDRY ROOM	5 LIVING ROOM	8 HOME BAR	11 BATHROOM
3	KITCHEN	6 TOILET	9 BEDROOM	12 STUDY

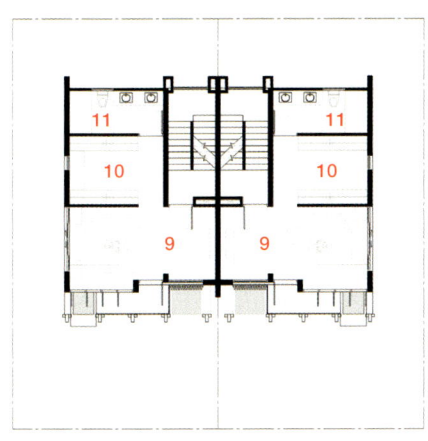

2ND FLOOR PLAN

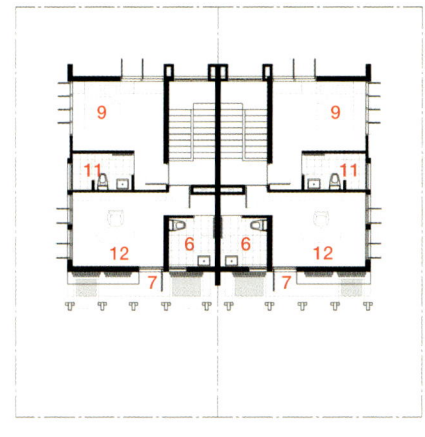

3RD FLOOR PLAN

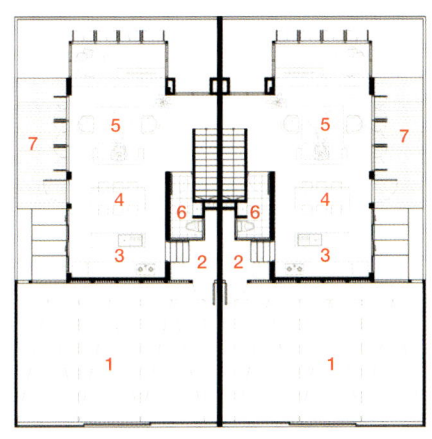

1ST FLOOR PLAN

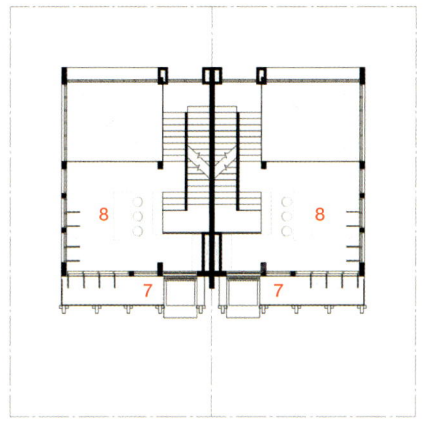

MEZZANINE FLOOR PLAN

레지던스 콩카허브

RESIDENCE KONGKAHERB

ARCHITECT : V2IN ARCHITECTS / THANAWIN PATTANAWONG

THIS PROJECT WAS STARTED ACCORDING TO plant expansion of the Kongkaherb that is needed to expand size of the new home office in order to cover increased number of staff. Also, the owner intended to build new home for the member of next generation.

From the part of the site that main part is plant construction which is an asymmetric land including angles that cannot be utilized efficiently. Also, the limited space which is challenge to design the functions to have garden area. The project architect decided to design all functions separately regarding size specification and divided into many boxes. Then he blended each part into every single angle site, putted all parts as one box and expanded each function independently in order to utilize all set back spaces and free spaces of building efficiently.

From the exterior of the building it looks like a simple white building. As the site was decorated by insulated glass in keeping the interior of the house simple, modern and tidy. Having the tree court in each floor in promoting of modern herb atmosphere. The building was designed according to simplicity and arranged inside functions independently. For mass and void were aligned with perceptions and direction which is reflected to building. The main design form of this project is "Cantilever" that is a provided a clear space underneath to make this building look outstanding and can be utilized every single space.

The remaining spaces were designed to be a courtyard for planting trees. There are openings around for air ventilation in order to bring sunlight inside building's rooms. The unused spaces were arranged to be garden which will diffuse and create an atmosphere around the building. There will be shade caused by the main area that is unfolded as a shade to help control appropriate light entering the building. Also, the effect of light and shadow during the day are created interesting functions inside the building. Starting from the 1st and 2nd floor is the office area. And the reception area of the residential area is the 3rd and 4th floor which is the privacy section for living. Moreover, the rooftop is relaxing space as the owner is friendly and familiar with the function that linked them to working space convincedly by walk down from their residence to meet with employees immediately. Thus, the combination of the office and the house is clearly arranged as private zoning but can be seen each other from every part of common are of the building through clear glass, courtyard, including the executive room that can be communicated to the staff by just opening the window.

Location Nakhon Pathom,Thailand **Use** Residential **Site area** 821.68 m² **Built area** 983m² **Gross area** 983m² **Project manager** Thanawin Pattanawong **Contractor** ARTCON CO., LTD. **Photographer** Rungkit Charoenwat

← Exterior & main gate   → Exterior view

↑ Panoramic view   ↙ Night view   ↳ Exterior side garage view

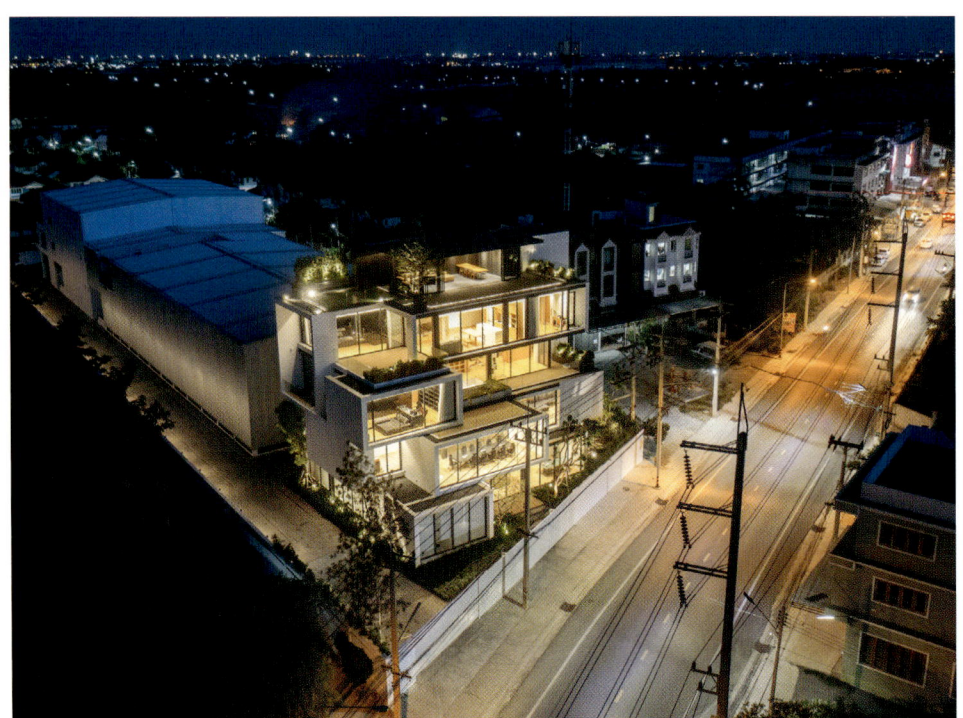

← Street view   → Exterior view

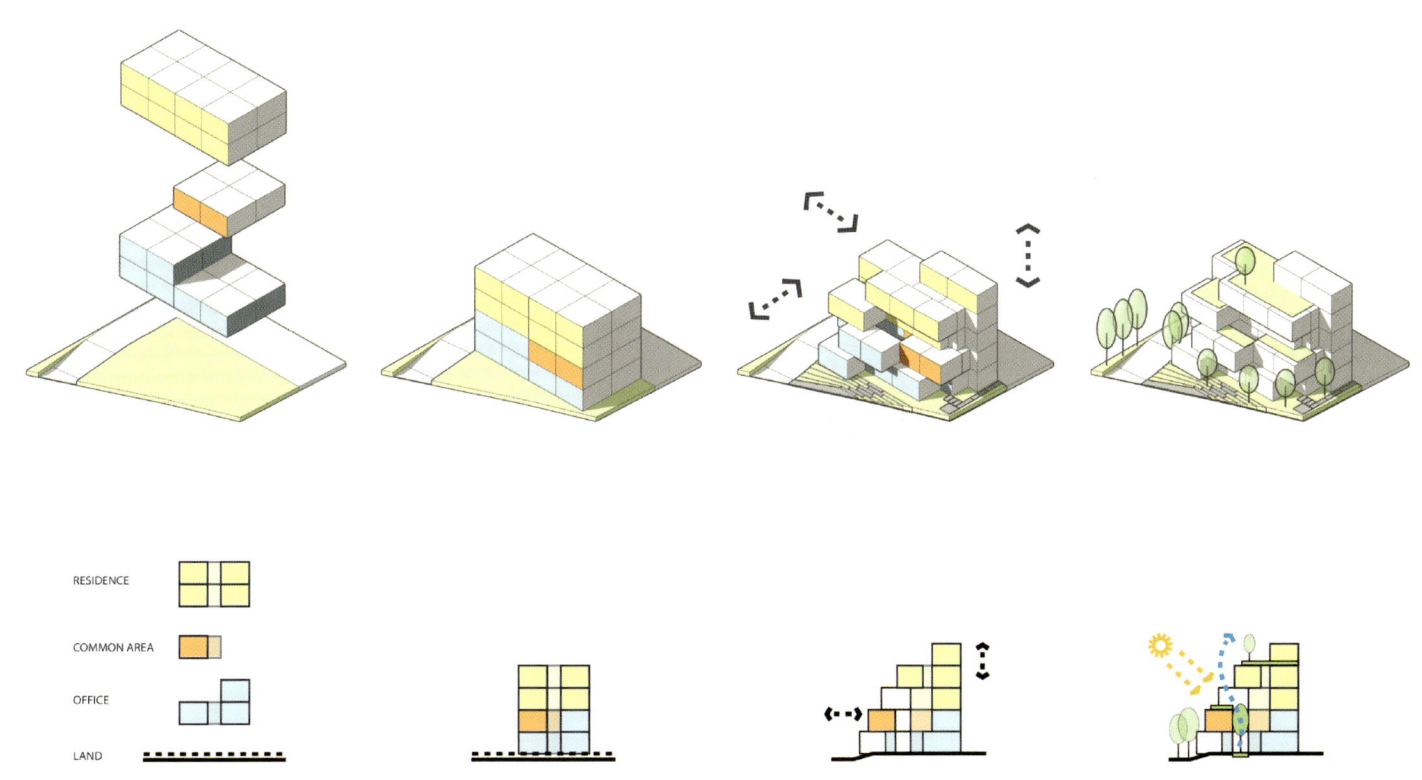

DIAGRAM

이 프로젝트는 콩카허브의 공장 확장에 따라 시작되었으며, 증가한 직원을 수용하기 위해 새 본사 사무실의 크기를 확장할 필요가 있었다. 또한, 소유주는 차세대 구성원을 위한 새로운 주택을 건설하고자 하였다.

대지의 주요 부분은 비효율적으로 활용될 수 없는 각도를 포함하는 비대칭적인 토지로 구성된 플랜트 건설이다. 또한, 제한된 공간은 정원 구역을 설계하는데 도전이었다. 건축가는 크기 사양에 따라 모든 기능을 별도로 설계하고 여러 상자로 나누고, 각각의 부분을 사이트의 모든 단일 각도와 혼합하여, 모든 상자를 하나로 두고 건물의 모든 후퇴 공간과 자유 공간을 효율적으로 활용하기 위해 각 기능을 독립적으로 확장하였다. 건물의 외관은 단순한 흰색 건물처럼 보였다. 내부를 단순하고 현대적이며 깔끔하게 유지하기 위해 단열 유리로 사이트를 꾸몄다. 각 층의 현대적인 허브 분위기를 촉진하는 나무 뜰을 배치하였다. 건물은 단순함에 따라 설계되었으며 내부 기능은 독립적으로 배열되었다. 질량과 공간은 건물에 반영된 인식과 방향에 맞춰 조정되었다. 이 프로젝트의 주된 디자인 형태는 "캔틸레버"로, 그 아래 명확한 공간을 제공하여 건물이 돋보이게 하고 모든 공간을 활용할 수 있게 하였다. 남겨진 공간들은 나무를 심기 위한 뜰로 설계되었다. 건물 내부공간에 햇빛을 끌어들이기 위해, 주변에는 공기 순환을 위한 개구부가 마련되었다. 사용되지 않는 공간들은 건물 주변의 분위기를 조성하고 정원으로 조성되었다. 건물에 적절한 빛을 조절하는 데 도움이 되는 그늘은 주요 영역에 의해 만들어졌다. 또한, 하루 중 빛과 그림자의 효과는 건물 내부에 흥미로운 기능을 창출하였다. 1층과 2층은 사무 공간이며, 주거 지역의 리셉션 공간은 사생활 보호 섹션으로서 3층과 4층에 위치한다. 또한, 소유주에게 옥상은 휴식 공간이면서 거주지와 작업 공간의 직원들을 만나러 갈 수 있는 연결 기능으로 익숙함과 친숙함을 제공하였다. 따라서 사무실과 주택의 조합은 개인적인 구역으로 명확하게 배열되어 있지만, 건물의 공용 구역의 모든 부분에서 투명한 유리를 통해 서로 볼 수 있으며, 집무실을 포함한 뜰은 창문만 열면 직원들과 소통할 수 있다.

↑ Each floor garden & veranda view

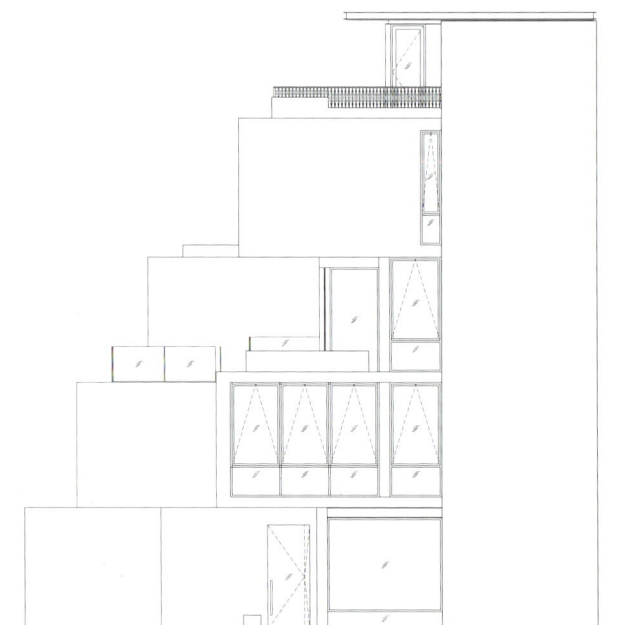

ELEVATION

↑ Decorated by insulated glass   ↵ Front view   ↳ 4th & roof floor residence view

↑ Garden & 1st floor terrace

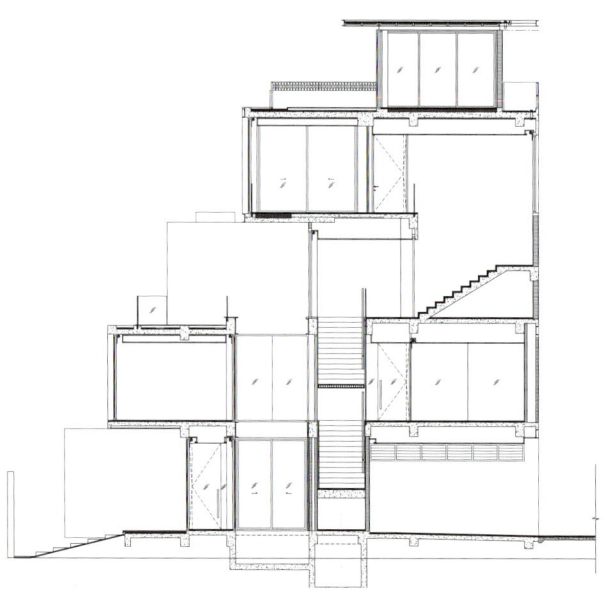
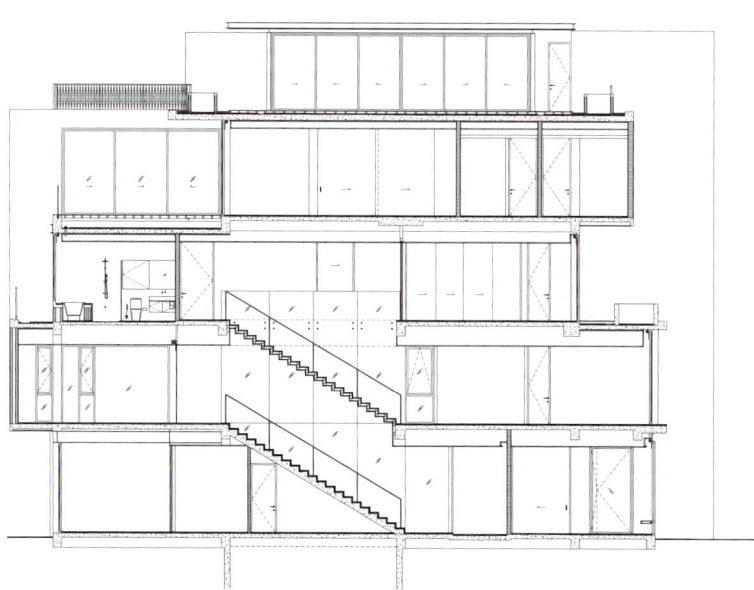

SECTION

↗ Garden & 1st floor terrace
→ 1st floor corridor & office
↘ 1st floor corridor

↑ 2nd floor courtyard   ↰ 2nd floor stair   ↱ 2nd floor stiar & dining area   ↳ 2nd floor kitchen area   ↳ 2nd floor kitchen & dining area

↰ Stair system   ↲ 3rd floor corridor   ↳ Courtyard

1. FIRE ESCAPE
2. LIFT HALL
3. FOYER
4. GARAGE
5. LIVING ROOM
6. TOILET
7. COURTYARD
8. POND
9. TERRACE
10. CORRIDOR
11. MEETING ROOM
12. OFFICE
13. STORAGE
14. PANTRY
15. MAIN GATE
16. ROAD
17. GARDEN

1ST FLOOR PLAN

← 4th floor kitchen & dining area   → 4th floor kitchen & dining area   ↳ Room   ↳ Toilet

1 FIRE ESCAPE	6 KITCHEN AREA	11 MD. ROOM	16 TERRACE
2 LIFT HALL	7 DINING ROOM	12 MEETING ROOM	17 ROOF GARDEN
3 TOILET	8 LIVING ROOM	13 FITNESS	18 WORK IN CLOSET
4 SERVICE ROOM	9 COURTYARD	14 RELAX AREA	19 MULTIPURPUSE ROOM
5 LAUNDRAY ROOM	10 CORRIDOR	15 BEDROOM	

2ND FLOOR PLAN

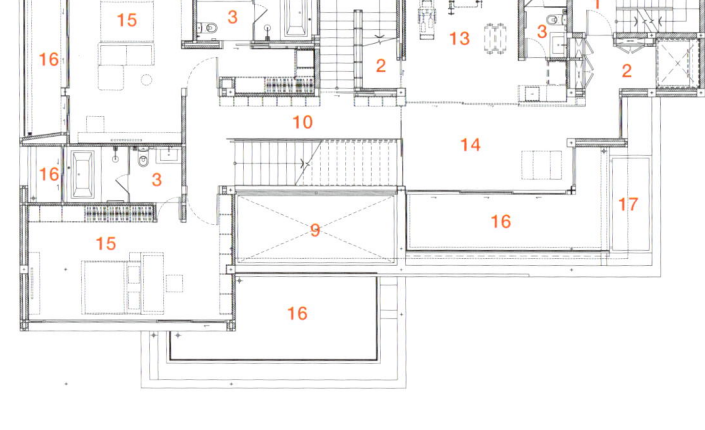

3RD FLOOR PLAN

← 4th floor terrace　↑ Roof floor multupurpose room & terrace　↳ Roof floor terrace

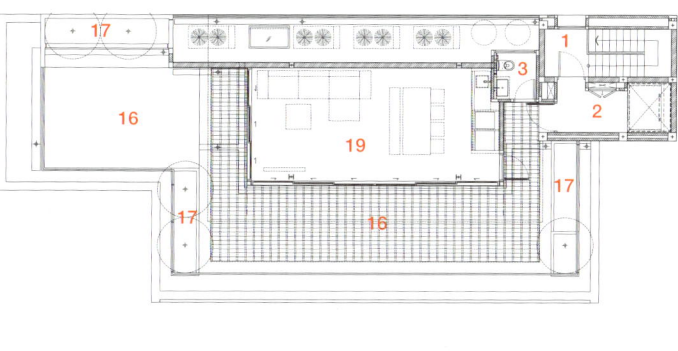

4TH FLOOR PLAN　　　　　　　　　　　　　**ROOF FLOOR PLAN**

타일 네스트
TILE NEST

ARCHITECT : H&P ARCHITECTS / DOAN THANH HA

THE HOUSE IS LOCATED IN A NEW DEVELOPING URBAN AREA IN Phu Ly city (Ha Nam province). Since this area is supposed to have graves deep underground (due to its proximity to the former cemetery), it is, therefore, necessary to excavate the old soil layer before construction so as to clear the above-mentioned assumption. This perspective then helps develop the concept of making use of the excavated site as part of the house (after the old soil layer is removed) so that not only distinguishable spacial features are created but also the possibility to harness geothermal energy is made. Named as Tile Nest, the house communicates the idea of creating a space, a blend of the Nest with many nooks and crannies finding all their ways up to the ground and the Ancient Pit House partially hidden underground. This combination gives the house's architecture a distinct corrugated appearance, with the shell felt like porous / perforated on the outside and large space on the inside. The outer shell is made up of many tiles suspended as if they were flying (provoking a thought of a stacked roof, and a sunshade, as found in a traditional house manner); The middle layer is characterized with transitional green balconies at different heights, which brings about captivating views and helps regulate the microclimate for spaces for use inside.

For a long time, tile has become a familiar and popular material with Vietnamese people, yet it is applied to this house in an unusual way to make its presence felt by seeing through, touching and sensing properties, thereby creating different but close experiences in the space of flower-like patterns from sunshine reflection,, wind and scent of plants - an Architecture immersed in nature realized by a full-of-memory personality

Location Phu Ly city, Ha Nam province, Vietnam **Use** House **Site area** 300m² **Total floor area** 450m² **Project manager** Doan Thanh Ha **Design team** Doan Thanh Ha, Nguyen Hai Hue, Trinh Thanh Huyen, Luong Thi Ngoc Lan, Tran Van Duong, Ho Manh Cuong, Nguyen Van Thinh **Photographer** Le Minh Hoang

이 주택은 하남성 푸리시의 새로 개발되는 도시 지역에 있다. 이 지역이 과거 묘지에 가까워 땅속 깊숙이 묘지가 있을 것으로 추정되므로 이를 해소하기 위해 건설 전 오래된 토양층을 굴착하는 것이 필요했다. 이러한 관점은 발굴된 부지를 주택의 일부로 활용하는 개념을 개발하는 데 도움이 되며(오래된 토양층이 제거된 후), 이는 뚜렷한 공간적 특성을 창조하는 것뿐만 아니라 지열 에너지를 활용할 수 있는 가능성도 만들어 냈다. 타일 네스트라 명명된 이 주택은 많은 틈새와 구석이 지면까지 자신의 길을 찾는 '둥지'와 부분적으로 땅속에 숨겨진 '고대 구덩이 주택'과의 혼합된 공간을 창조하는 아이디어에서 출발했다. 이 조합은 주택의 건축에 독특한 뚜렷한 주름진 외관을 제공하며, 외관은 다공성 / 천공된 듯한 느낌을 주고 내부에는 넓은 공간을 갖추고 있다. 외부 껍질은 날고 있는 듯 매달린 많은 타일로 구성되어 있어, (전통 가옥의 방식에서 볼 수 있는 쌓인 지붕과 태양 가리개를 연상시킨다); 중간층은 다양한 높이의 녹색 발코니로 특징되며, 이는 매혹적인 전망을 제공할 뿐만 아니라 내부 사용 공간의 미기후를 조절하는 데 도움을 주었다.

오랜 시간 동안 타일은 베트남 사람들에게 친숙하고 인기 있는 재료가 되었지만, 이 주택에는 특별한 방식으로 적용되어 투시하고, 접촉하며, 물성을 감지함으로써 그 존재감을 느낄 수 있게 했다. 이는 태양빛 반사로 만들어진 꽃 모양 패턴의 공간에서 바람과 식물의 향기로 다양하지만, 가까운 경험을 창출한다. 기억에 가득 찬 개성으로 자연에 묻혀 있는 건축을 실현한 것이다.

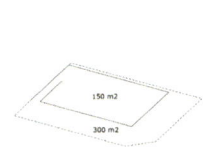 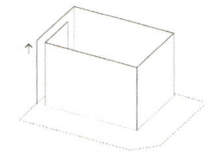 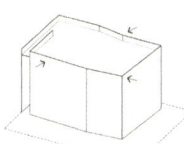 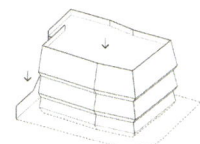

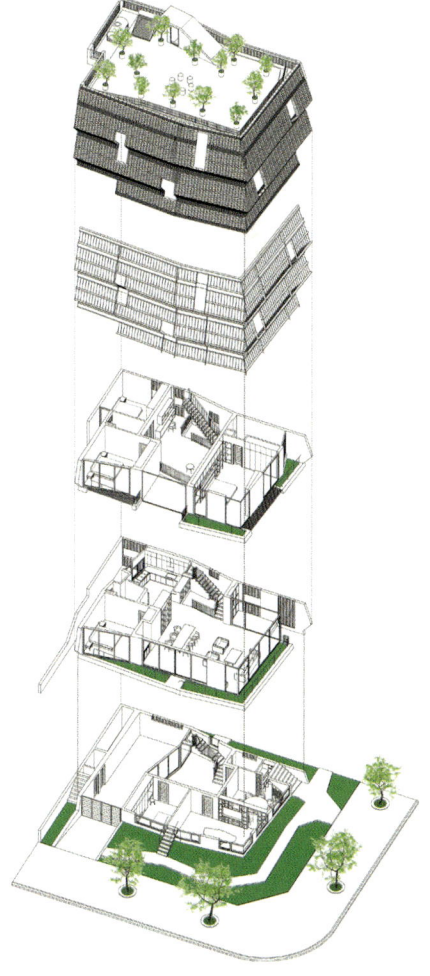

DIAGRAMS

← Exterior view

↑ South-east view   ↙ Exterior & site night view

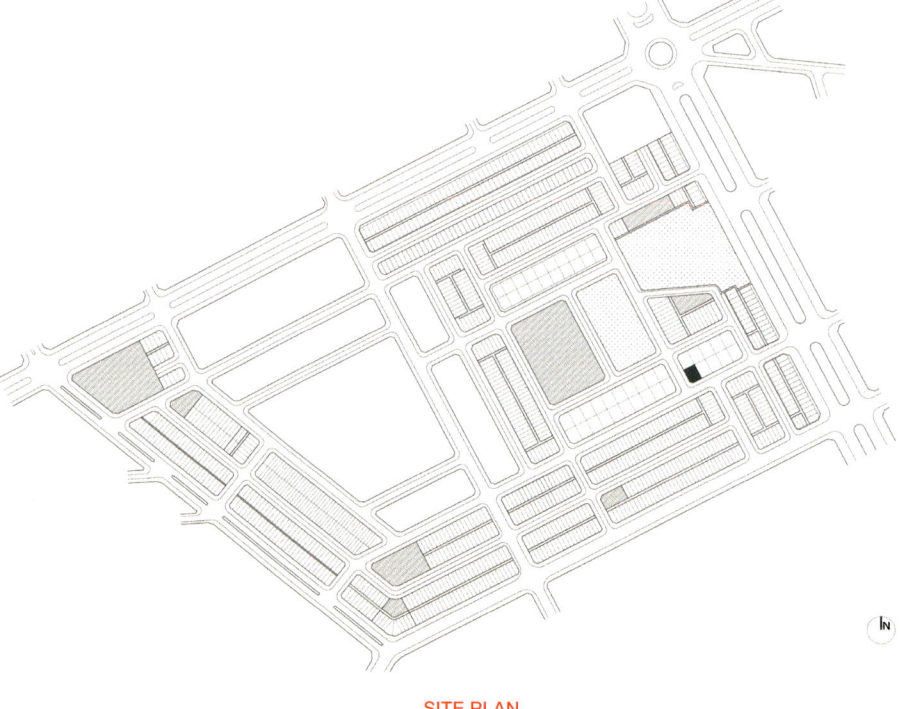

SITE PLAN

↑ Exterior & street view

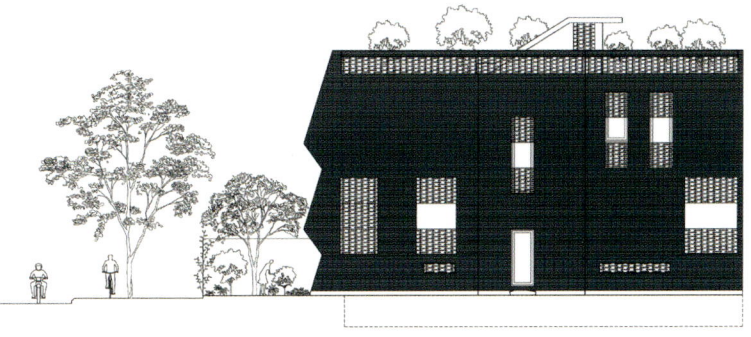

WEST ELEVATION

SOUTH ELEVATION

EAST ELEVATION

SECTION

↑ South exterior view   ← North-east exterior view   → South-west exterior view

The nest with many nooks and crannies finding all their ways up to the ground   ♡   The ancient pit house   →   Tile nest

CONCEPT SKETCH

← Facade tile detail → Facade tile detail ↙ Facade tile detail ↳ Facade tile detail

↑ The outer shell is made up of many tiles

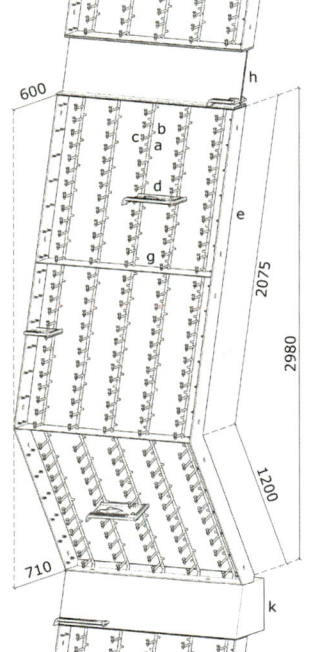

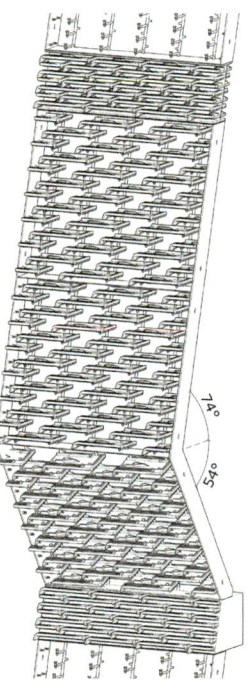

Typical units
a. reinforced steel bars, d=12mm
b. reinforced steel bars, d=6mm
c. bulong
d. roof tiles 22

e. plate steel (8mm)
g. plate steel (5mm)
h. I-beam (h=350mm)
k. concrete beam (h=350mm)

roof tiles 22
(size: 340X200X12mm)

drill holes+create slots

TILES DETAIL

↑ Tiles facade & balcony　↙ Entrance　↓ Tiles facade & balcony　↳ Stair

201

↑ 1st floor living room   ← 1st floor living & lobby   ⌐ Ground floor multifunctional space   ↳ 1st floor living room & window system

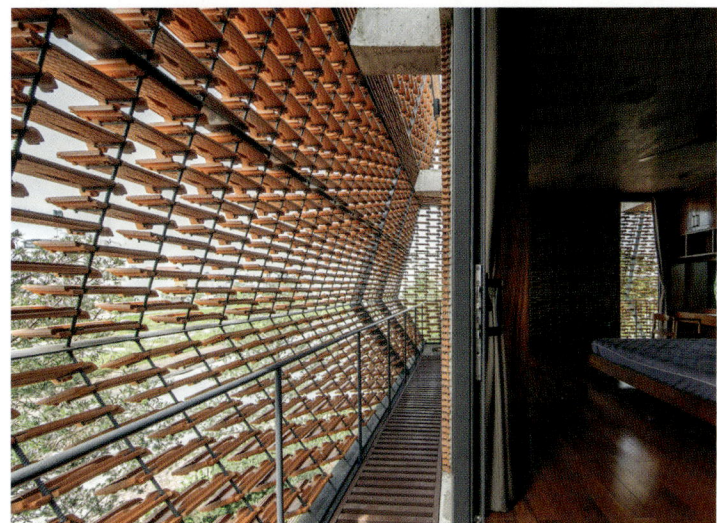

← 2nd floor multifunctional space   → Multi-use room   ↲ Bedroom   ↳ Bedroom & balcony

1. ENTRANCE
2. GARAGE
3. MULTIFUNCTIONAL SPACE
4. BEDROOM
5. MULTI-USE ROOM
6. GARDEN
7. TOILET
8. STORAGE
9. WATER TANK
10. TECHNICAL ROOM
11. LOBBY
12. LIVING ROOM
13. DINING ROOM
14. KITCHEN
15. WASHING ROOM
16. WORSHIP
17. STUDY & WORK SPACE
18. VOID
19. BALCONY
20. ROOFTOP GARDEN

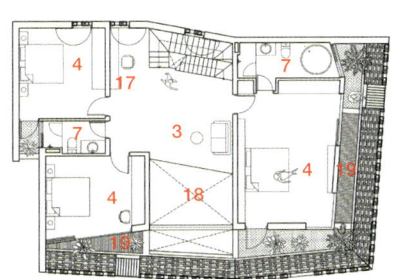

2ND FLOOR PLAN

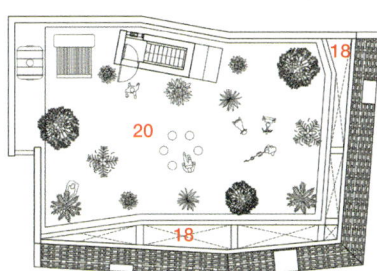

ROOF TOP PLAN

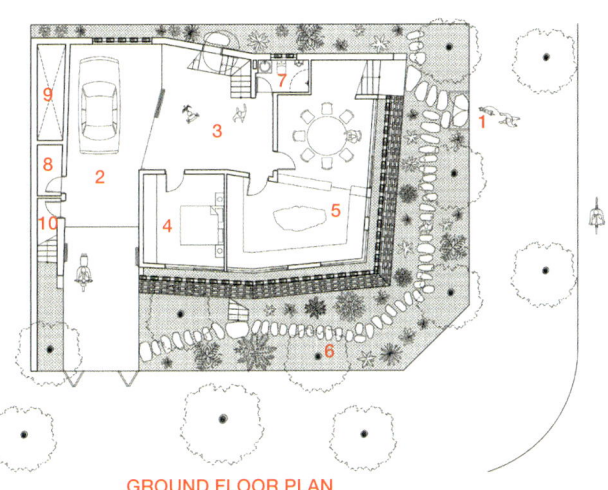

GROUND FLOOR PLAN

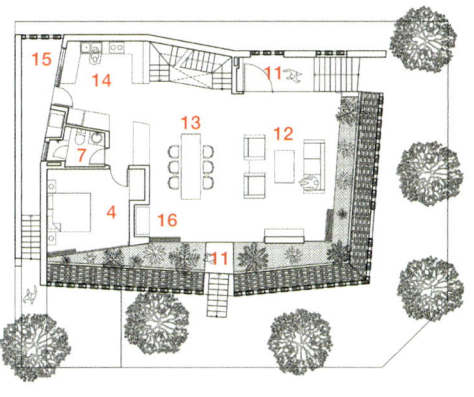

1ST FLOOR PLAN

하우스 벨라쿠
HOUSE BELAKU

ARCHITECT : 4SITE ARCHITECTS / CHANDRAKANT KANTHIGAVI

BELAKU, WHICH MEANS "HOUSE OF LIGHT," is a residential project on the outskirts of Bengaluru. A 2BHK tenant house and a duplex home were requested by our client. The home needed to be well-lit and to have natural light throughout. Both the project's cost and Vaastu Shastra compliance must be acceptable.
We began by planning to let in a lot of light by cutting wide openings on the north and east sides. We have a finite amount of "Stack Effect" ventilation openings. We have made the project future-proof by giving 4 feet wide room for lighting and ventilation on the entire north face of the building, keeping in mind the potential tenants on the constrained urban plot. The main door should be on the NE corner, the Master Bedroom on the SW, and the kitchen on the SE, according to traditional Indian "Vaastu Shastra."
The main room in the house, the Brahmasthan Living and Dining, has double height. The double-height space connects to a landscape garden, with the balcony garden's security envelope extended outside so that curtain creepers can cover it. The SW corner master bedroom that connects to a private garden is where our clients stay.
The flooring was maintained dark with black leather, polished black stone, and a splash of yellow granite to reflect the natural light. Family area flows over the dining area, fostering a sense of unity within the family and tying together the client's sons' two bedrooms. The bedroom in the SW corner opens to the north garden. The front bedroom on the east side should also have a balcony access. For all of the master bedrooms' garden balconies, we have a brick screen wall. Brick Screen wall completely encloses the north and east of the bedroom to provide privacy.
The purpose of the C-shaped envelope is to imply a connection between the two residences. The purpose of the Brick Jali (perforated screen), which lets diffused light into the area but prevents rainwater from entering, was to make the building look cohesive.

Location KNS Unnati, Komagatta Road, Sulikere, Karnataka, India **Use** Residential **Site area** 110m² **Built area** 290m² **Gross area** 383m² **Completion** 2022 **Project manager** Veeresh Mutnal **Design team** Aishwarya Nainegali, Ramya Joshi, Avinesh, Rachana **Contractor** Stone Edge Construction **Photographer** Studio Recall

"빛의 집"을 의미하는 벨라쿠는 벵갈루루 외곽의 주거 프로젝트이다. 고객으로부터 2BHK 임대 주택과 복층 주택이 요청되었다. 해당 주택은 조명이 밝아야 하며 자연광이 내부 전체에 퍼져야 했다. 프로젝트 비용과 "Vaastu Shastra" 준수 여부 둘 다 요청 되었다.
북쪽과 동쪽 면에 넓은 개구부를 만들어 많은 빛을 들일 수 있도록 계획을 시작했다. 우리는 제한된 양의 "스택 효과(Stack Effect)" 환기 개구부를 갖고 있다. 도심 플롯의 제약을 염두에 두고 잠재적인 임차인을 위해 건물의 전체 북쪽 면에 너비 4피트의 조명과 환기를 위한 여유 공간을 두어 프로젝트를 미래 지향적으로 만들었다. 전통적인 인도 "Vaastu Shastra"에 따르면, 주 출입문은 북동쪽 모서리에 있어야 하며, 주 침실은 남서쪽에, 주방은 남동쪽에 위치해야 했다.
주택의 중심 공간인 브라마스탄 거실과 식당은 두 배의 높이를 가지고 있다. 이 두 배 높이 공간은 조경된 정원과 연결되며, 발코니 정원의 보안 울타리는 밖으로 확장되어 커튼 크리퍼 식물이 덮을 수 있도록 설계되었다. 남서쪽 모서리에 위치한 주 침실은 개인 정원과 연결되며, 여기에 클라이언트가 머문다. 바닥재는 천연광을 반사할 수 있도록 검은색 가죽, 광택 있는 검은색 석재, 그리고 노란색 화강암의 포인트로 어두운 톤을 유지하였다. 가족 공간은 식당으로 자연스럽게 이어지며, 가족 간의 일체감을 증진하고 클라이언트의 두 아들의 침실을 연결한다. 남서쪽 모서리에 위치한 침실은 북쪽 정원으로 통하며 동쪽에 있는 전면 침실 역시 발코니 접근이 가능하다. 모든 주 침실의 정원 발코니에는 벽돌 스크린 벽을 설치했다. 벽돌 스크린 벽은 침실의 북쪽과 동쪽을 완전히 둘러싸며 개인 프라이버시를 제공한다. C자 형태의 외피는 두 거주 공간 사이의 연결을 암시하기 위한 목적으로 계획되었다. 브릭 잘리(천공 스크린)의 목적은 흩어진 빛은 내부로 들이고 비는 막으면서 건물의 조화로운 외관을 만든다.

← Main view → Night view

↑ Bird's eye view   ← South exteior view

SITE PLAN

↑ Panorama ↙ North exterior view

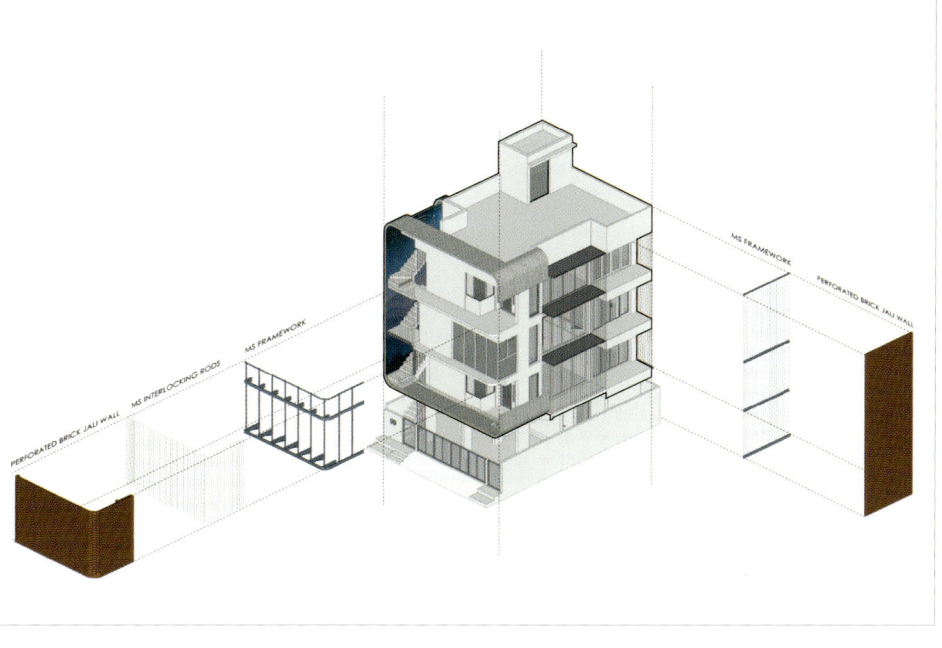

ISOMETRIC

207

← Main gate view   → East elevation detail

1 PARKING
2 KITCHEN
3 BEDROOM
4 LIVING ROOM
5 VERANDA
6 TERRACE

LONGITUDINAL SECTION   **CROSS SECTION**

← Facade brick detail ↑ Facade & window ↗ Brick detail

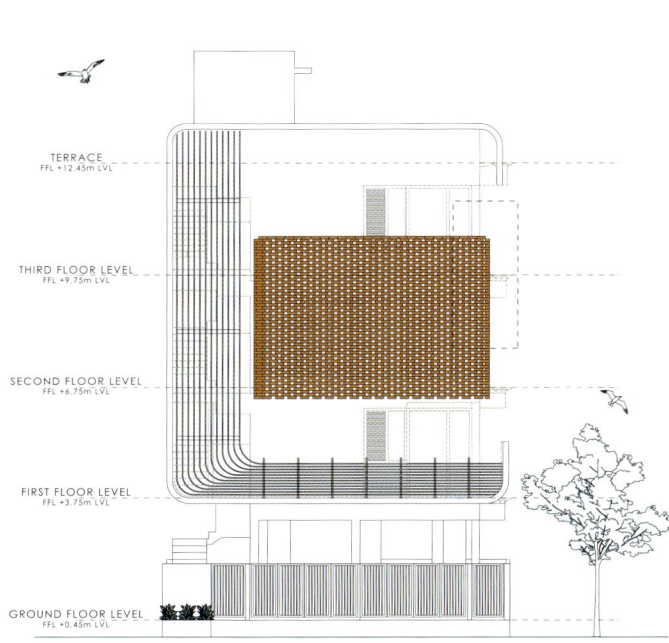

1. HANDRAIL
2. GRANITE COPING
3. L-ANGLE 100X100X10mm
4. 50X50X8mm HORIZONTAL MS MEMBER
5. 12mm DIA MS BRITE RODS AS PER INTERLOCK SYSTEM INSERYED INTO THE PERFORATED BRICKS
6. EXPOSED PERFORATED BRICK WASS AS PER COURSE DETAILS
7. EXISTING RCC SLAB
8. 150X75mm BOX SECTION
9. L-ANGLE 100X100C10mm
10. ISMB 150 HORIZON MS MEMBER (SUPPORTING CANTILEVER SLAB) @EVERY 977mm ANCHORED TO BEAM AND SLAB AS PER STRUCTURAL DETAIL
11. 50X50X8mm VERTICAL MS MEMBER(FRAMEWORK)@ EVERY 977mm ANCHORED AT TOP AND BOTTOM OF THE SLAB

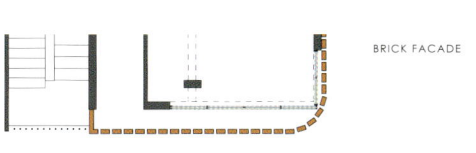

EAST ELEVATION

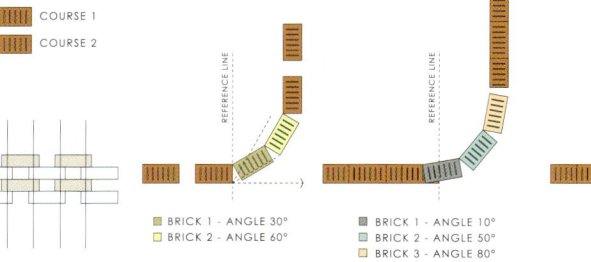

DETAILS

↑ 1st floor living room ↵ Stair → Veranda

↑ 1st floor living room & stair

1 VERANDA	4 KITCHEN	7 TOILET	10 FAMILY SPACE
2 LIVING ROOM	5 BEDROOM	8 SITOUT	11 MASTER BEDROOM
3 DINING ROOM	6 DRESSING ROOM	9 SPILLOUT	

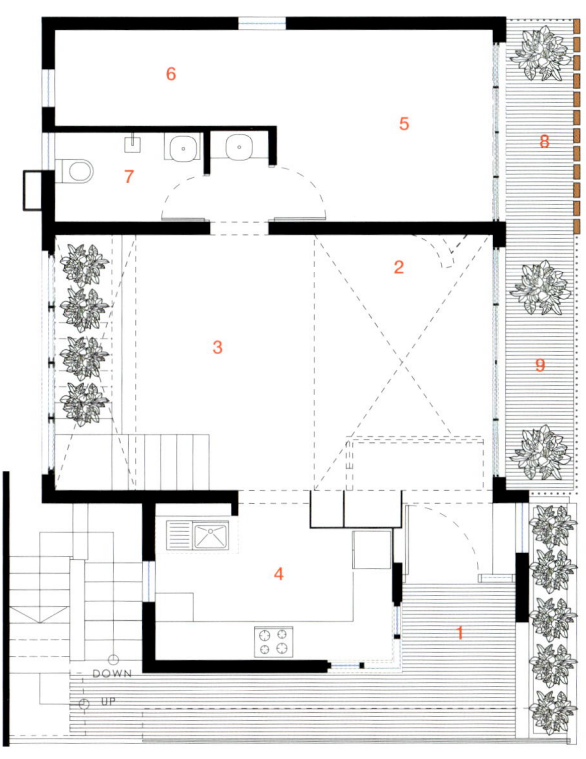

1ST FLOOR PLAN

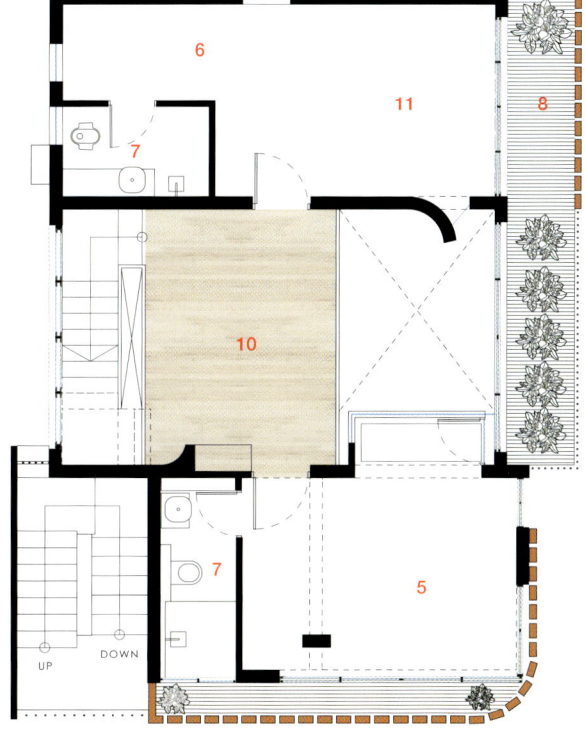

2ND FLOOR PLAN

비밀의 정원

SECRET GARDEN

ARCHITECT : ROOM+ DESIGN & BUILD / VINH PHUC TA

SINCE ITS FIRST APPEARANCE IN THE LATE 18TH CENTURY, townhouse has gradually become the most popular type of private houses in Vietnam. People living in townhouses take the advantages of convenient traffic, proximity to amenities, business opportunities, and quickly-rising property value over time. However, they always concern about the issues of security, safety and privacy as well as the lack of day lighting, natural ventilation and greenery space.

A 4-level townhouse located in a 95 square-meter corner site at the intersection between two alleys without sidewalk in Ho Chi Minh City was designed and built to adapt to the needs of a 4-people family, who asked for a contemporary and pleasant living space with the addition of some gardens and an elevator for easier movement. The project brief was fulfilled by two relating design solutions: optimizing the spatial layout within a built-form that responds to the site and needs; and proposing a unique arrangement of vibrant garden spaces.

Firstly, an optimized geometric order was proposed to thoughtfully locate the functional spaces while sensitively responding to the characteristics of the site, especially its irregular shape and round corner. A gentle set-back from the site corner gives space for the subtle entry and a leafy courtyard. The convex curving granite steps and the concave curving main door form an unusually elliptical verandah on the ground floor. The corner space on the first level is accommodated by the master bedroom's convex curving window and planter underneath an elliptical terrace on the second floor. When observed from outside, the play of convex and concave curves as well as the recessed glazing with deep planters create a sculptural built-form which enhances privacy, day lighting and natural ventilation.

In contrast with the relatively discreet look of the exterior facade, the interior space is dominated by transparency and openness. As one comes inside, he or she would surprisingly find a charming "secret garden" with a big acerola cherry tree and diverse types of plants hiding at the furthest corner of the site. The garden brings soft daylight and cool breeze into the heart of the house while also providing protection and privacy thanks to the decorative hollow-brick walls. A high-end hydraulic glass elevator and two internal balconies are placed aside the secret garden. Timber floor, beige painted walls as well as curving built-in cabinets create an elegant yet dynamic interior.

Location Ho Chi Minh City, Vietnam **Use** Residential **Site area** 95m² **Building area** 70m² **Gross floor area** 270m² **Completion** 2023 **Project manager** Anh Tuan Pham **Design team** Vinh Phuc Ta, Anh Tuan Pham, Minh Hieu Huynh, Bao Loc Hoang, Van Kieu Pham, Tan Trung Ta, Duc Truong Nguyen, Ngoc Vien Pham **Contractor** ROOM+ Design & Build **Photographer** Sonmeo Nguyen Art Studio

SKETCH

← Exterior view

← Street view   → Street view

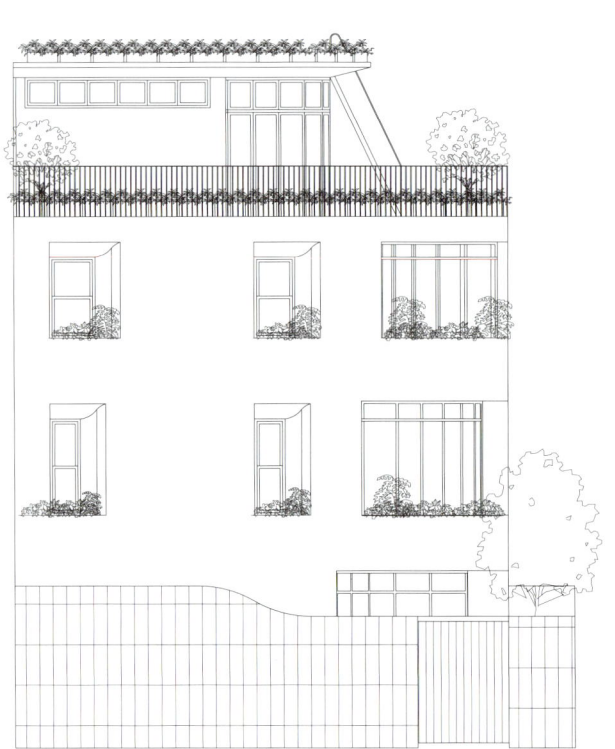
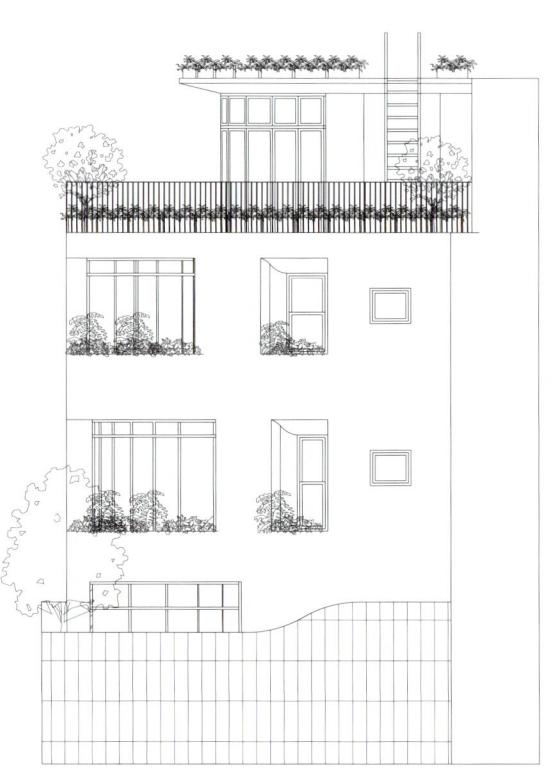

ELEVATION

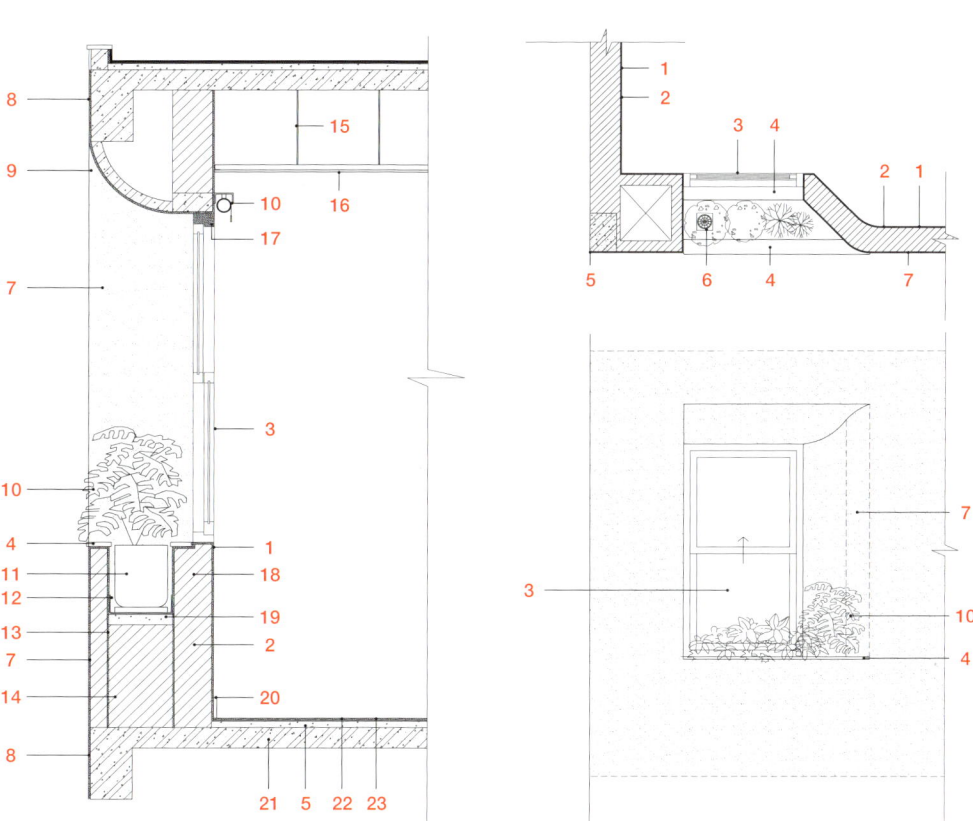

↖ Looking-up view   ↙ Partly view of exterior   ↑ Entrance

1. 10mm MORTAR LAYER + INTERIOR PAINT FINISH
2. 80mm BRICK WALL
3. ALUMINUM FRAME, 10mm TEMPERED GLASS WINDOW
4. 20mm THICK GRANITE SLAB WITH CURVED EDGES
5. REINFORCED CONCRETE COLUMN
6. STAINLESS STEEL FLOOR DRAIN AND PVC DOWNPIPE
7. 10mm MORTAR LAYER + EXTERIOR SAND-WASH PAINT FINISH
8. REINFORCED CONCRETE BEAM
9. 50mm THICK CURVING CONCRETE SLAB ON IRON MESH
10. WALL-MOUNTED WINDOW ROLLER BLIND
11. W250*L600*H300 PVC PLANTER POT
12. 1mm THICK BITUM MEMBRANE WATERPROOF LAYER
13. 25mm THICK VERSICELL DRAINAGE PANEL
14. BROKEN BRICK FILLED IN PLANTER BOX
15. IRON HANGING CABLE
16. 10mm THICK PLASTERBOARD ON ALUMINUM FRAME
17. REINFORCED CONCRETE LINTEL BEAM ABOVE WINDOW
18. 0.5mm FABRIC FILTER LAYER
19. 50mm CONCRETE SLAB
20. 10mm LAMINATED TIMBER PANEL
21. 100mm REINFORCED CONCRETE SLAB
22. 30mm MORTAL LAYER
23. 3mm PLASTIC LAYER

DETAIL

⌐ Lobby
— Interior view
⌐ Veranda

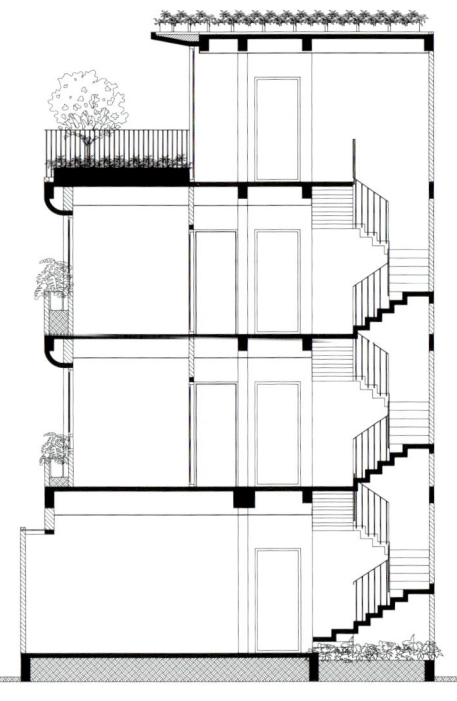

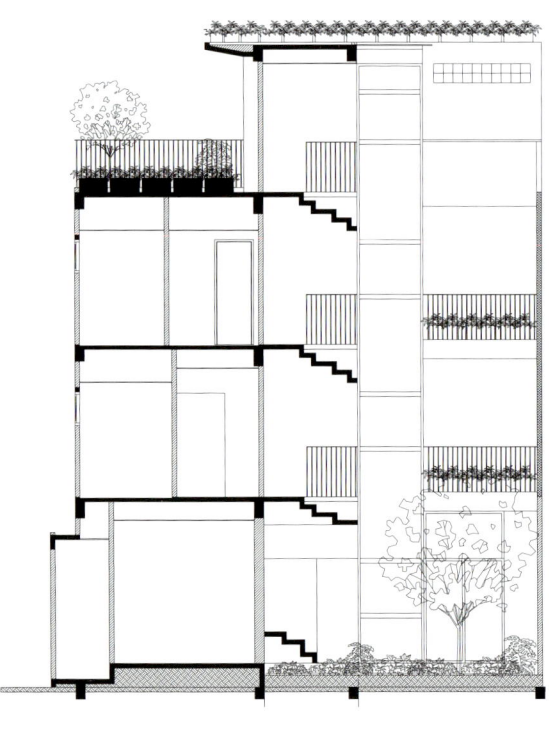

SECTION

↑ Interior view   ← Kitchen   → Secret garden

18세기 말 처음 등장한 이래로 타운하우스는 베트남에서 가장 인기 있는 개인 주택 유형이 되었다. 타운하우스에 사는 사람들은 편리한 교통, 편의 시설에 대한 접근성, 비즈니스 기회 및 시간이 지남에 따라 빠르게 상승하는 부동산 가치의 이점을 누리고 있다. 그러나 그들은 항상 보안, 안전 및 사생활 문제뿐만 아니라 일조, 자연 환기 및 녹지 공간 부족에 대해 우려한다.

호치민 시의 두 골목 사이, 인도가 없는 95㎡의 모퉁이 부지에 위치한 4층 규모의 타운하우스는 편리한 이동을 위해 정원과 엘리베이터를 추가하여 현대적이고 쾌적한 생활 공간을 원하는 4인 가족의 요구에 맞게 설계되었다. 프로젝트 요구 사항은 대지와 니즈에 대응하는 건축 형태 내에서 공간 레이아웃을 최적화하고, 활기찬 정원 공간의 독특한 배치를 제안하는 두 가지 관련 설계 솔루션을 통해 완성되었다.

먼저, 기능적 공간을 세심하게 배치하면서 대지의 특성과 불규칙한 형태와 둥근 모서리에 민감하게 반응하며 최적화된 기하학적 순서를 제안하였다. 대지 모퉁이의 부드러운 후퇴는 은은한 입구와 잎이 무성한 안뜰을 제공한다. 볼록한 곡선의 화강암 계단과 오목한 곡선의 주출입구는 지상층에 특이한 타원형 베란다를 형성한다. 1층 코너 공간은 2층 타원형 테라스 아래에 있는 안방의 볼록한 곡선형 창문과 플랜터로 수용되어 있다. 외부에서 관찰하면 볼록하고 오목한 곡선의 연주와 깊은 플랜터로 후퇴된 유리는 사생활 보호와 일조 및 자연 환기를 향상시키는 조각적 건축 형태를 만들어낸다.

비교적 신중한 외관 파사드와는 대조적으로, 내부 공간은 투명성과 개방성에 의해 지배된다. 내부로 들어서면, 놀랍게도 가장 안쪽 모퉁이에 숨겨진 매력적인 '비밀의 정원'과 큰 아세롤라 체리 나무와 다채로운 식물들을 발견하게 된다. 이 정원은 부드러운 채광과 시원한 바람을 집의 중심부로 가져오는 동시에 장식용 중공 벽돌 벽에 의한 보호와 사생활을 제공한다. 고급스러운 유압 유리 엘리베이터와 두 개의 내부 발코니가 비밀의 정원 옆에 배치되었다. 목재 바닥과 베이지색 벽 페인트와 배치된 곡선형 맞춤 캐비닛은 우아하면서도 역동적인 인테리어를 완성한다.

↑ Void   ↵ Staircase   ↵ Glass lift

← Rooftop garden   → Rooftop garden

1 ENTRANCE	6 LIVING ROOM	11 TOILET	16 BALCONY
2 COURTYARD	7 DINING ROOM	12 MASTER BEDROOM	17 BEDROOM
3 BIKE & BICYCLE PARKING LOT	8 KITCHEN	13 WALK-IN CLOSET	18 TERRACE
4 VERANDA	9 GLASS LIFT	14 BATHROOM	19 ROOFTOP GARDEN
5 LOBBY	10 SECRET GARDEN	15 FAMILY ROOM	20 LAUNDRY & STORAGE

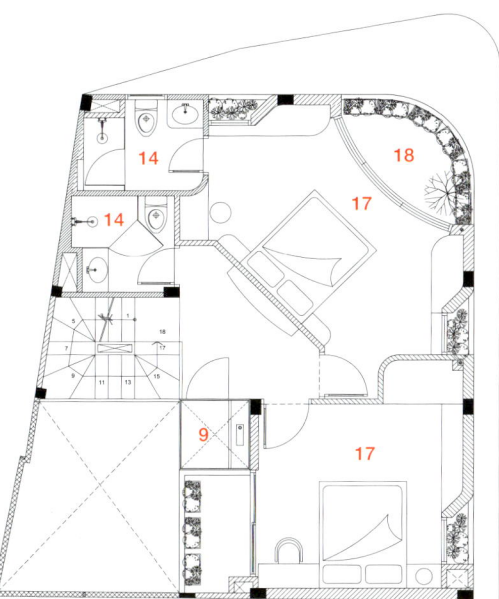

2ND FLOOR PLAN

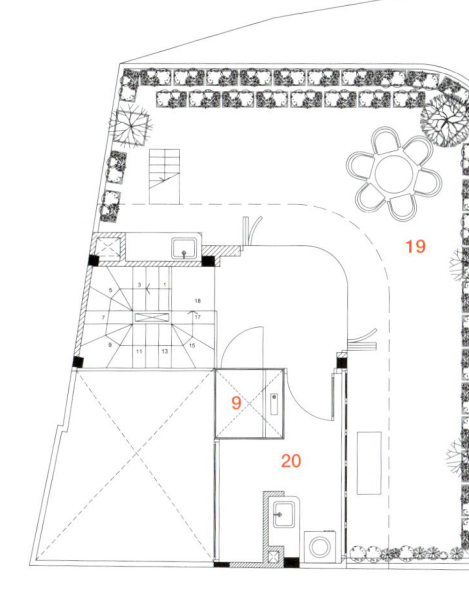

3RD FLOOR PLAN

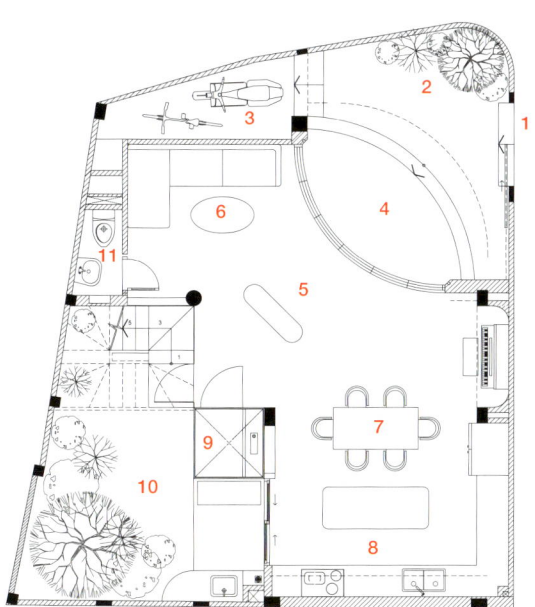

GROUND FLOOR PLAN

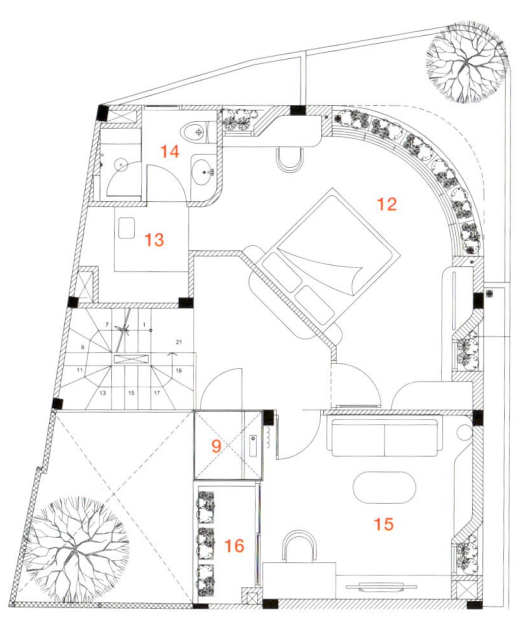

1ST FLOOR PLAN

HOUSE WITH AN EYE

눈이 있는 집

ARCHITECT : STUDIO ANNA JACH GMBH / ANNA JACH

Location Zurich, Switzerland **Use** Housing **Gross floor area** 155m² **Completion** 2023 **Art** Oskar Zieta (steel canopy), Drzach&Suchy (Eye Facade Relief) **Photographer** Alexandre Kapellos

CLIENTS BRIEF & PLANNING CONSTRAINTS

Extension of the entry area, new guest's bathroom, two-story library for 500 books with a swing for the youngest child, yoga & meditation room, child's bedroom, rooftop terrace.

Very narrow but high frequented street, reduced parking possibility for the contractors, the client wanted to move back in as soon as possible (short construction time).

Sustainability as the key-aspect of the building construction

Massive reduction of the carbon footprint by the usage of local wood and local insulation material (blown-in cellulose). Short transportation ways due to the selection of local contractors and materials. Excavation earth was reused as a rammed-earth sitting bench in the garden. Solar panels on the rooftop + heat pump provide electric energy and hot water. Due to the thickness of the building insulation the heating cost was reduced by 50%. Inside, wooden panels were left without adding an extra render as the final finishing surface, resulting in a natural way of humidity exchange, pleasant room climate and the smell of fresh pine wood. No artificial chemical treatments were used for the internal wall protection, the surfaces were impregnated with natural flax oil combined with 5% of white mineral paint to prevent the wood from getting the yellow stain by the time.

Digital art

Building facade with a programmed 3D shadow relief: CNC-wave-cut vertical facade planks cast a shadow of a closed eye in the morning which opens during the course of the day, giving a "living" effect to the exterior skin of the building and creating an interaction with the surrounding. By the artists Drzach&Suchy Blown-up steel canopy by the artist and product designer Oskar Zieta.

건축주 요청 및 계획의 제약

확장된 현관, 새로운 게스트 욕실, 막내아들을 위한 그네를 포함한 500권의 책을 보관할 수 있는 2층 규모의 서재, 요가 및 명상실, 아이 침실, 옥상 테라스.

매우 좁지만 통행량이 많은 도로, 계약업체를 위한 주차 공간 부족, 클라이언트는 가능한 한 빨리 입주하기를 원함. (짧은 공사 기간).

건물 건설의 핵심 측면으로서의 지속가능성

현지 목재와 현지 단열재(블로운인 섬유소)를 사용하여 탄소 발자국을 대폭 줄임. 현지 계약업체 및 자재 선정으로 인한 운송 거리 단축. 굴착한 흙을 정원의 앉을 수 있는 벤치로 재사용. 옥상의 태양열 패널 + 히트펌프로 전기에너지와 온수 공급. 건물 단열재의 두께로 인해 난방비 50% 절감. 내부는 최종 마감재로 별도의 렌더링을 하지 않고 목재 패널을 그대로 두어 자연스러운 습도 교환과 쾌적한 실내 기후, 신선한 소나무 냄새를 느낄 수 있게 함. 내부 벽을 보호하기 위해 인공적인 화학 처리를 하지 않았으며, 시간이 지나면서 목재에 노란색 얼룩이 생기는 것을 방지하기 위해 표면에 천연 아마 오일과 5%의 흰색 광물성 페인트를 혼합하여 함침시킴.

디지털 아트

프로그래밍된 3D 그림자 부조가 있는 건물 외관 : CNC 웨이브 커팅된 수직 파사드 판자는 아침에 감은 눈의 그림자를 드리우다가 낮이 되면 열리면서 건물 외피에 '살아있는' 효과를 부여하고 주변 환경과 상호작용을 일으킴. 아티스트 드르자크&수취(Drzach&Suchy) 제공. 예술가이자 제품 디자이너인 오스카 지에타(Oskar Zieta)의 블로우업 강철 캐노피.

← Exterior view   → Night view

← Exterior view   → Exterior view   ↙ Partly view

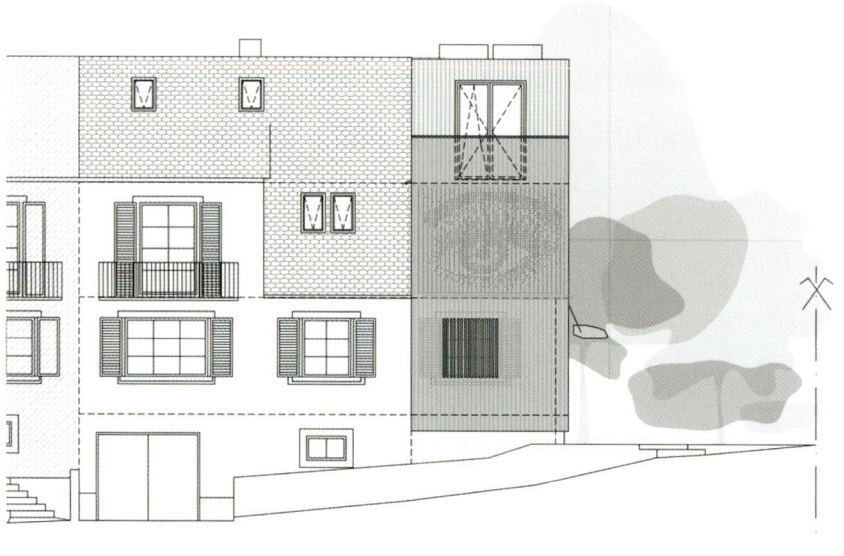

SIDE ELEVATION

↑ Looking-up vieww   ↙ View of the sunset   ↘ View of the sunset

223

↑ Dining room   ← Foyer   ↳ Stair

⌐ Living room
└, Living room

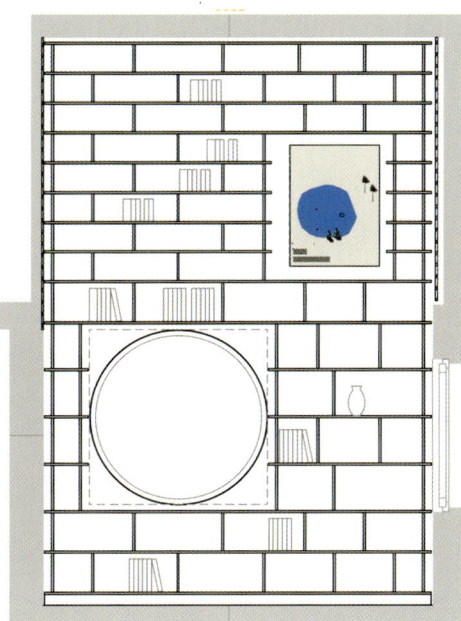

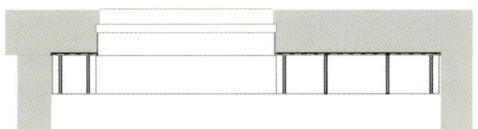

DETAIL OF THE LIBRARY

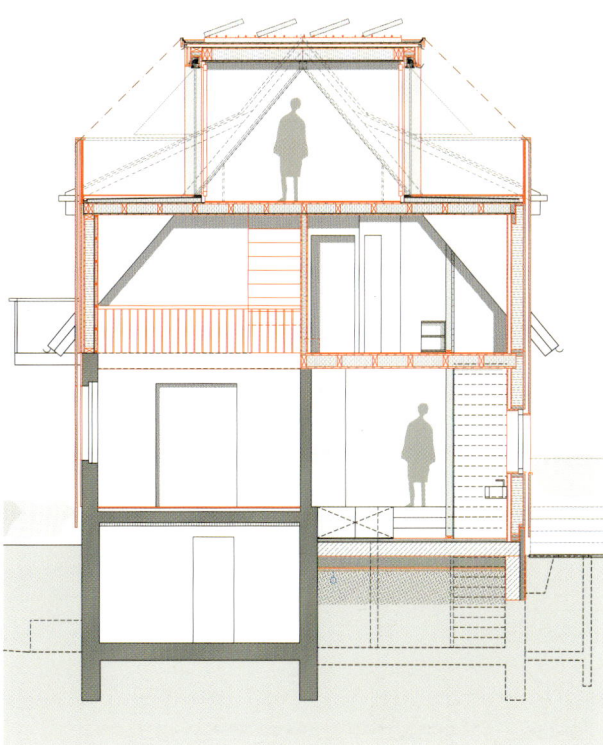

SECTION

← Interior view   → Study

1 FOYER
2 KITCHEN & DINING ROOM
3 LIVING ROOM
4 STUDY
5 BATHROOM
6 ROOM
7 BEDROOM
8 TERRACE

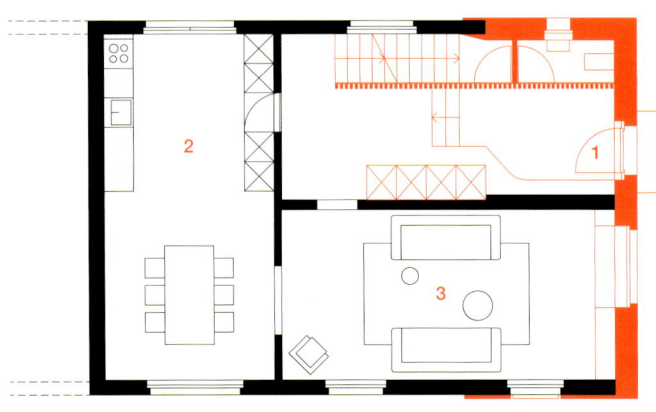

1ST FLOOR PLAN

2ND FLOOR PLAN

↑ Room on the 3rd floor   ↳ Interior view on the 2nd floor

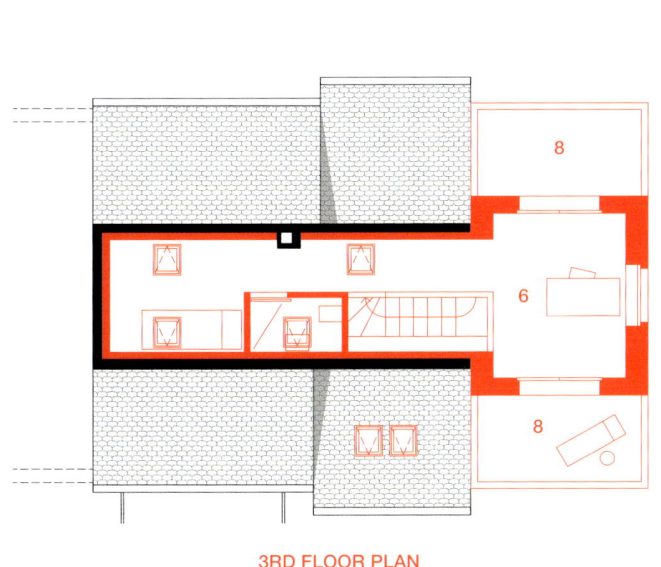

3RD FLOOR PLAN

ISAGI RESTAURANT

ARCHITECT : ATELIER SUPERB + CHEN-TIEN CHU ARCHITECT / ERIC YU + CHEN-TIEN CHU

THE WEST DISTRICT OF TAICHUNG IS A POPULAR RESIDENTIAL and artistic hub of the city, it was one of the earliest historical communities established for the US army. As the land value soars, high-rise developments start to thrive nearby. These clusters of two-storey duplexes are the only area where the streetscape was still preserved.

Isagi restaurant sits in one of the townhouse duplexes. Previously used as an artist's studio, the existing building was painted in pale yellow and light green. As this new restaurant becomes one of the occupiers in the life cycle of this building, we wish to not completely erasing the traces it has accumulated over time, and to avoid presenting the restaurant space in an entirely new language. We endeavor to seek for a harmonious conversation between the new and the obsolete. We believe this aligns with the owner's pursuit of creating an innovative and wholesome Japanese catering experience by serving dishes of chef's choice.

For floor plan layout, the service space of the restaurant is concealed and stacked vertically upstairs. It maximises the view towards the garden for the dining area. The reconfiguration of the linear stairs divides the space more efficiently. The four new load-bearing walls not only partition the kitchen, the front serving and prep area, the private dining rooms, and at the same time serve as the structural reinforcement. On the upper floor, the large north-facing private dining room can be separated into two small private rooms if required. The dining space facing the west is designated to accommodate smaller groups of guests, with the boxed wooden seats at the back acting like a gazebo encouraging the guests to look out into the garden.

The new window frames of this renovated building are deliberately offset from the external walls. The sharp-edged metallic finish aesthetically contrasts with the rawness of exposed concrete walls. Layered industrial-grade plywood sheets were used in a certain portion of the ceilings. Mediating a cascading relationship between the original structure and the new space, that deductively induces new perspectives between the present and past.

Location Taichung City, Taiwan **Use** Restaurant **Site area** 238m² **Gross floor area** 198m² **Built area** 99m² **Project manager & design team** Eric Yu, Chen-Tien Chu **Contractor** Chen-Tien Chu **Photographer** Studio Millspace

← Exterior view   → Entrance view

↑ Exterior night view

SECTION

↑ Garden view   ↙ 1st floor kitchen & bar   ↳ Entrance

타이중 서쪽지역은 인기 있는 주거 및 예술 중심지로, 미군을 위한 최초의 역사적 공동체 중 하나로 자리 잡았다. 부동산 가치가 치솟으면서, 고층 개발이 인근에서 번성하기 시작했다. 이 2층 복층 주택군은 거리 경관이 여전히 보존된 유일한 지역이다.

이사기 레스토랑은 타운하우스 복층 건물 중 한 곳에 자리 잡고 있다. 이전에는 예술가의 작업실로 사용되던 기존 건물은 연한 노랑과 밝은 녹색으로 칠해져 있었다. 이 새로운 레스토랑이 건물의 수명 속에서 한 자리를 차지하게 되면서, 우리는 시간이 쌓아 올린 흔적을 완전히 지우지 않으며, 레스토랑 공간을 새로운 언어로 제시하는 것을 피하고자 했다. 우리는 새로움과 낡은 것 사이의 조화로운 대화를 모색하려고 노력하였다. 이는 셰프의 선택으로 요리를 제공함으로써 혁신적이고 온전한 일식 경험을 창조하려는 소유주의 추구와 일치한다고 믿었다.

레스토랑의 평면 배치에 있어, 서비스 공간은 위층으로 숨겨져 수직적으로 쌓였다. 이는 식사 공간에서 정원을 향한 전망을 극대화한다. 선형 계단의 재배치는 공간을 더 효율적으로 나눈다. 새로운 4개의 하중 벽은 부엌, 앞쪽 서빙 및 준비 공간, 개인 다이닝 룸을 분리할 뿐만 아니라, 동시에 구조적 보강 역할도 한다. 상층에는 큰 북쪽을 마주하는 개인 다이닝 룸이 필요에 따라 두 개의 작은 개인 룸으로 분할될 수 있다. 서쪽을 마주하는 식사 공간은 소규모의 손님들을 위해 마련되었으며, 뒤쪽에 배치된 박스형 목재 좌석은 정자처럼 손님들이 정원을 바라보도록 하였다.

이 개조된 건물의 새 창문틀은 외벽에서 일부러 비켜 나 있다. 날카로운 모서리의 금속 마감은 노출된 콘크리트 벽의 거친 느낌과 미적 대비를 이룬다. 일부 천장에는 산업용 등급의 합판이 층층이 사용되었다. 기존 구조와 새 공간 사이에 계단식 관계를 매개하면서, 현재와 과거 사이에 새로운 관점을 유도하는 공간을 창출하였다.

↑ 1st floor garden & hall view   ↙ 1st floor bar

SECTION

← 1st floor dining area   → Japanese style bar   ↙ 1st floor stair   ↘ 1st floor dining area & stair

SECTION

↑ 2nd floor dining area   ↓ 2nd floor dining area

← Small size dining area   → 2nd floor staircase

1 MAIN GATE
2 ENTRANCE
3 HALL
4 DINING AREA
5 JAPANESE STYLE BAR
6 KITCHEN
7 TOILET
8 STAIR
9 SMALL SIZE DINING AREA

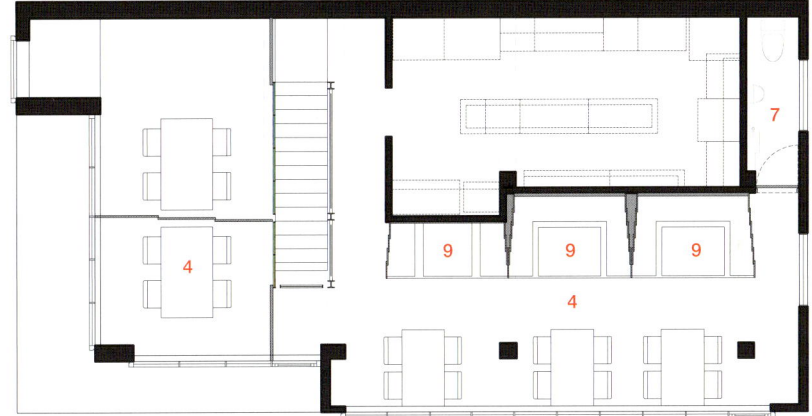

2ND FLOOR PLAN

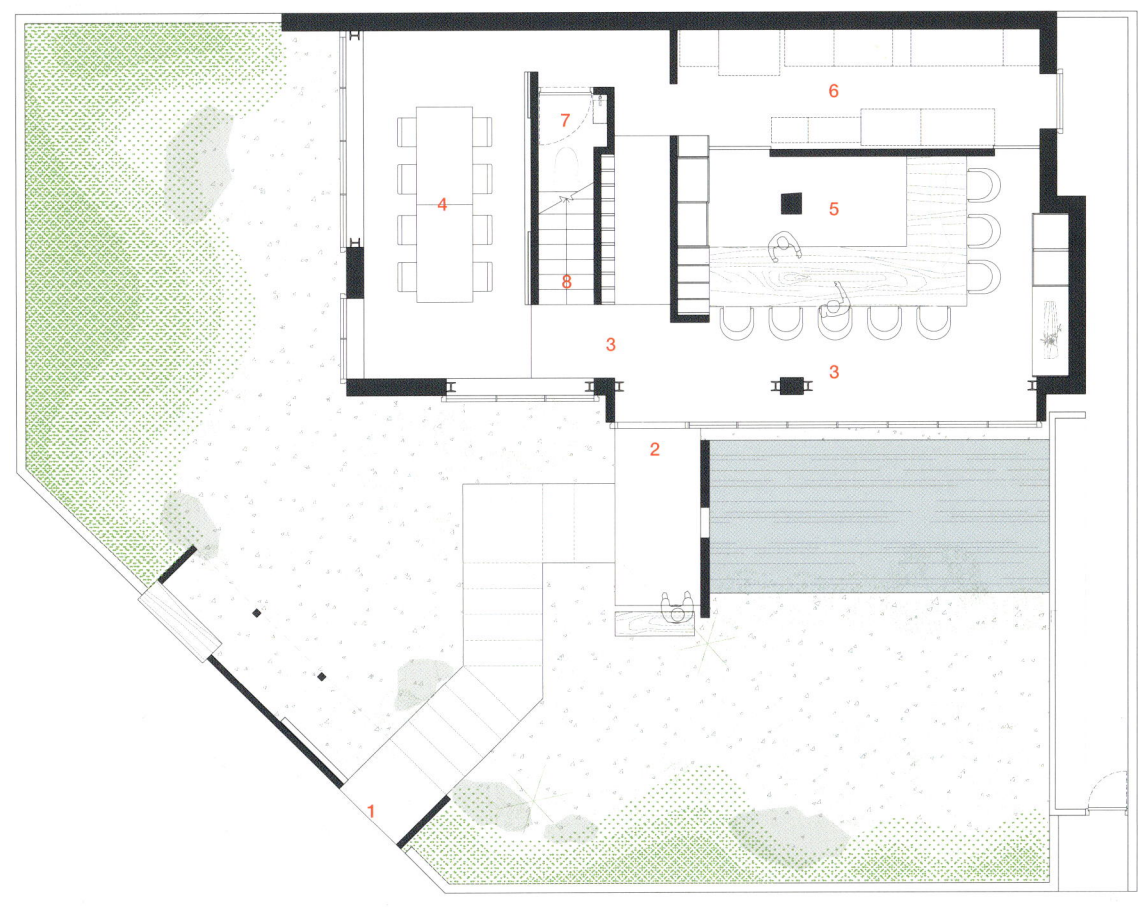

1ST FLOOR PLAN

액솔로틀
THE AXOLOTL

ARCHITECT : YU2E / BILL TSUI

THIS IS A PROJECT DEVELOPED UNDER THE TOC (transit oriented communities) standards. It is intended to spur multi-family development near public transit by relaxing restrictions like required yards, parking and density maximums or floor area in exchange for setting aside some of those units for lower income households and tenants.

Our design approach was to maximize the opportunities afforded by the TOC incentives to create more livable dwellings at a more affordable costs. Reducing parking requirements eliminates the costs of a subterranean level. Additional benefits cascade through the building allowing dwelling units on the ground floor and larger private open spaces like balconies and roof decks for the upper floors. Larger exterior spaces combined with large floor to ceiling sliding doors allow for more outdoor utility and generous natural light and ventilation inside. Tactical distribution of these decks throughout also reduced the overall bulk and scale of the building minimizing impact to the street level and adjacent neighbors.

While a unifying central courtyard was not tenable in this arrangement, the common access corridor is designed to create a sense of community and shared space by utilizing alternating panels and openings in lieu of a typical dark and uninviting double loaded corridor. The units are efficiently organized into the limits of a type V three story building at grade with a single stair core, which keeps costs lower while also making possible the kind of open layout units and multiple wall openings and windows on all sides.

With characteristics taken from both bungalow courts and dingbat apartments, our project is an evolutionary hybrid of Los Angeles middle housing. The dwelling units are at once cozy and livable, breezy with ample access to outdoors, while benefiting from the cost efficiency of multiple units collectively sharing spaces and resources to strengthen community.

Location Los Angeles, CA, USA **Use** Multi-family dwelling **Site area** 627m² **Building area** 650m² **Gross floor area** 836m² **Completion** 2023 **Project manager** Bill Tsui **Design team** Kimberly Lawes, Sascha Mohkami **Contractor** Giraffe LA **Photographer** Taiyo Watanabe

이 프로젝트는 TOC(대중교통 중심 커뮤니티) 기준에 따라 개발되었다. 이는 대중교통 근처에서 다세대 주택 개발을 촉진하기 위해 필수 마당, 주차 공간, 밀도 최대치 또는 층 면적에 대한 제한을 완화하는 대신 일부 주택을 저소득 가구와 세입자를 위해 할당하는 것을 목적으로 한다.

디자인 접근 방식은 TOC 인센티브가 제공하는 기회를 최대한 활용하여 더 저렴한 비용으로 더 살기 좋은 주거 공간을 창출하는 것이다. 주차 요구 사항을 줄임으로써 지하층의 비용이 감소되었다. 더불어 이는 건물 전체로 이어져 1층에 주거 단위와 상층에 발코니와 옥상 데크와 같은 프라이빗한 외부공간을 확보할 수 있었다. 넓은 외부 공간과 바닥부터 천장까지 이르는 슬라이딩 문을 통해 실외 활용도를 높이고 내부에 풍부한 자연광과 환기를 제공한다. 이러한 데크의 전술적 분배는 건물의 전체적인 부피와 규모를 최소화하여 가로 높이와 인접한 이웃에 미치는 영향을 최소화하였다.

이 배치에서 통일된 중앙 안뜰은 불가능했지만, 공용 복도는 전형적인 어둡고 이중적인 복도 대신 패널과 개구부를 번갈아 사용함으로써 공동체감과 공유공간을 조성하도록 설계하였다. V형 3층 건물은 계단코어 하나로 V형 3층 건물의 한계로 효율적으로 구성되어 있어 비용을 절감하는 동시에 개방형 배치 유닛과 여러 개의 벽면 개구 및 창문을 활용할 수 있다.

방갈로 코트와 딩바트 아파트의 특징을 모두 갖춘 우리의 프로젝트는 로스앤젤레스 미들 주택의 진화적 혼합체이다. 주거 단위는 아늑하고 살기 좋으며, 넓은 실외 접근성과 함께 쾌적한 환경을 제공하며, 여러 단위가 공간과 자원을 공유하여 커뮤니티를 강화하는 비용 효율성의 이점을 누릴 수 있다.

← Exterior view

↑ Front facade

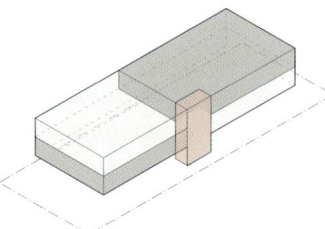

Typical quadplex under existing development standards where majority of the ground floor is reserved for parking

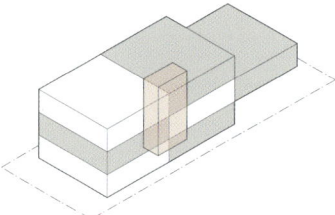

Toc standards allow for 7 dwelling units and reduction of required parking in exchange for a unit reserved for low income tenant

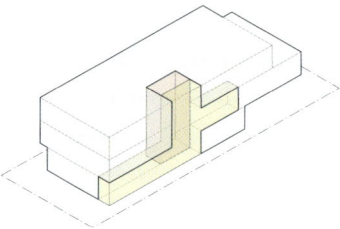

Carving out circulation core and access corridors

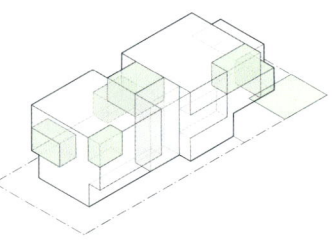

Carve out large exterior decks and private open spaces

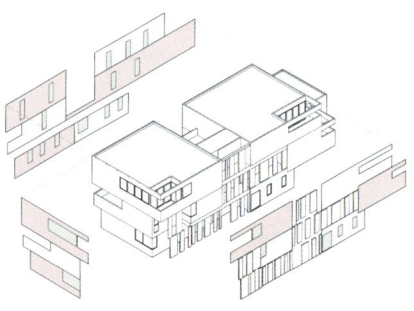

Finish exterior elevations as index of individual dwelling units

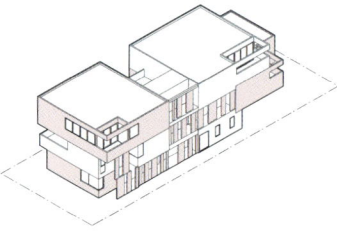

The Axolotl

DESIGN DIAGRAM

↑ Exterior view   ↵ Outdoor deck   ↳ Entrance

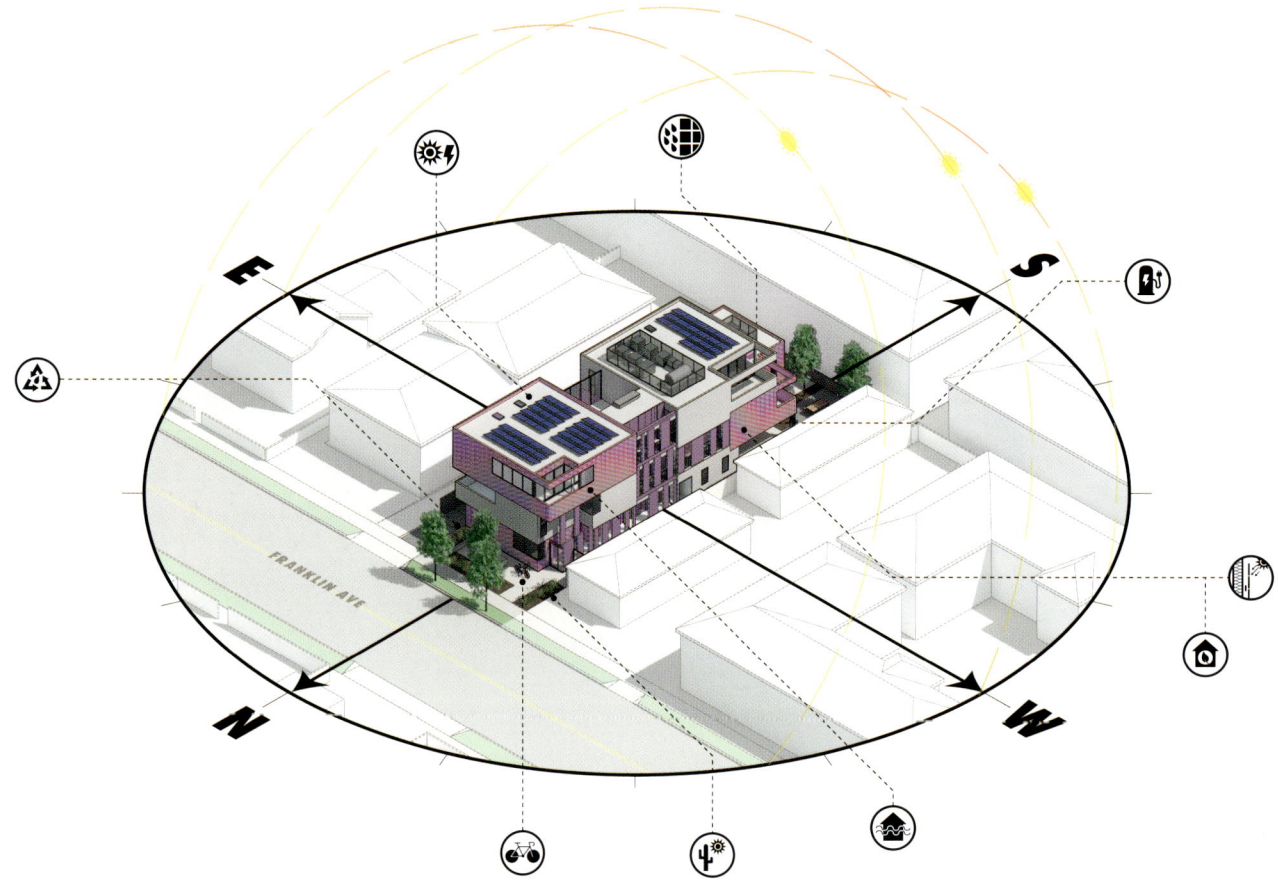

 Renewable PV Power:
allows for lot to be off the grid and supply energy back to the city in the summer months

 EV Charging Station:
plug-in friendly

 Residential Infill:
TOC approval allowed for 1.75x the amount of units in this RD1.5 zone

 Treat + Reuse Runoff for Irrigation:
Sub-grade cistern storage tank for collection of storm water runoff

 Sustainable Building Materials:
Fiber cement board is sustainably sourced, weather resistant, and long lasting

 Drought Tolerant Flora:
less water intensive plants implemented to reduce environmental impact

 Rain Screen:
exterior insulation + rain screen increase envelope performance

 Cross Ventilation:
operable windows + sliding glass doors allow for Increased air flow facilitating in passive cooling

 Permeable Paving:
reduces urban runoff and assists in groundwater recharge

 Transit Oriented Community:
with only 4 parking spots on-site, tenents are encouraged to use more sustainable modes of transportation

SUSTAINABILITY DIAGRAM

← Living room → Living room

↑ Living room & Deck

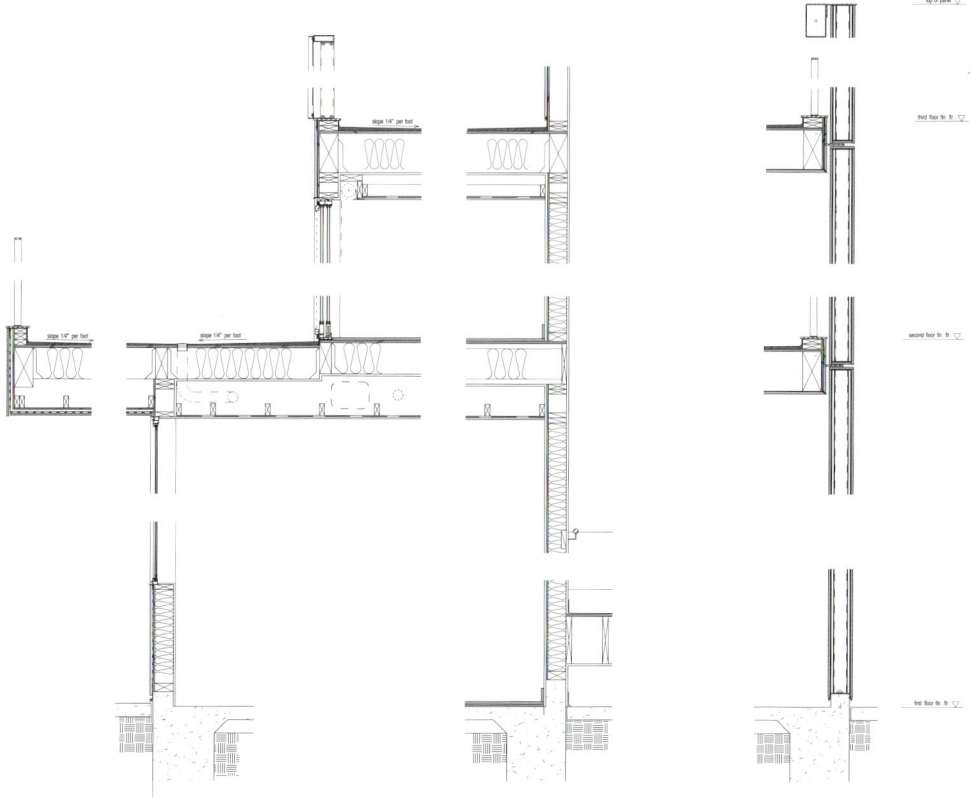

SECTION DETAIL

↑ Kitchen & Dining room

1	FRONT YARD	3	ENTRANCE	5	STAIRCASE	7	PARKING LOT
2	BICYCLE PARKING LOT	4	HALLWAY	6	DWELLING UNITS	8	BACKYARD

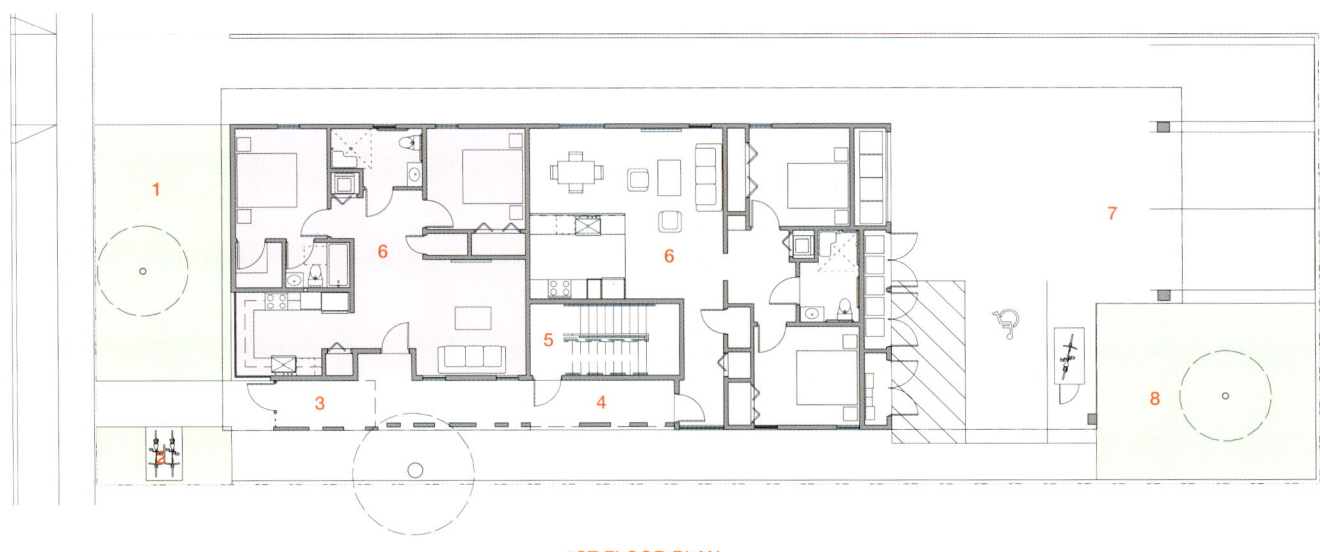

1ST FLOOR PLAN

← Interior view → Kitchen

3RD FLOOR PLAN

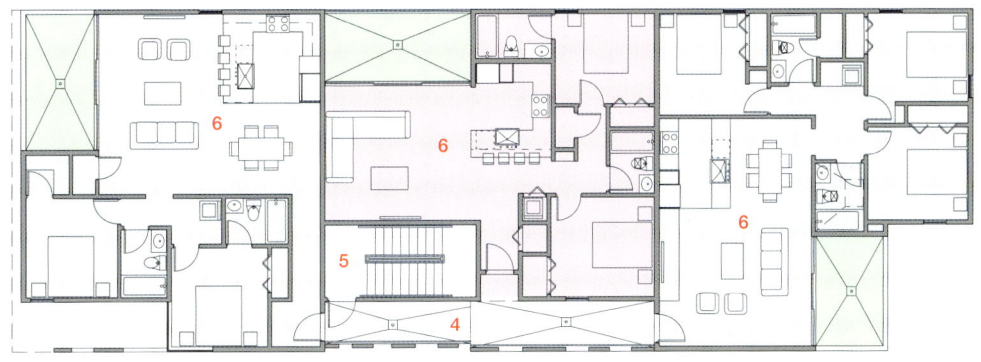

2ND FLOOR PLAN

하우스 커버
HOUSE COVE(R)

ARCHITECT : TOUCH ARCHITECT / SETTHAKARN YANGDERM, PARPIS LEELANIRAMOL

A TWO-STORY SINGLE-DETACHED HOUSE SETTLES in a village near Suvarnabhumi Airport, with its size of 150m², located in only two-third part of 256m² plot. It was too small for a family with five members; a couple with two kids and grandma. Not only small curve-triangular land shape, but also located at the corner which connects to both sides of public street. An extension shape is similar to a cove which parallel with the curve of the land, in order to use the space to its full potential. Three-floor is created to combine all functions needed while setback area is used for an outdoor tiny landscape with a relaxation pond.

Steel structure is applied, mainly for reducing beams' depth which affects total building height, yet it helps decreasing column size which increasing more interior space. Two staircases are placed in different corner which minimize circulation. Double space in between the stairs with skylight allows natural sunlight through downstairs, instead of creating vertical voids with lacks privacy.

As the house located at the corner while privacy is needed, facade cover is required. Vertical aluminum trellis together with perforated metal are used to cover all cove, continuing to an existing house, to harmonize both buildings together. Slanted facade shape is needed for the extension, as it has higher level than the previous, so the facade line is sloped down to link between the two. However, to cover the existing house, simple flat shape is enough for the existing one, thus it is unnecessary to create a non-parallel facade for this part. For the maintenance issue, powder-coated aluminum is used for the upper floors facade which is not only lightweight, but also zero-maintenance, since it is hard to fix or repaint at the high level. However, metal is used for the house wall, since it can easily be maintained, while cheaper price in Thailand.

Location Bangphli, Samutprakarn, Thailand **Use** Housing **Site area** 627m² **Gross floor area** 170m² **Completion** 2021 **Project manager** Setthakarn Yangderm, Parpis Leelaniramol **Design team** Pitchaya Tiyapitsanupaisan, Supanan Tangsaj januraksa, Tanita Panjawongroj **Engineering** Chittinat Wongmaneeprateep **Contractor** DWN Builder **Interior Contractor** TRIGON Construction **Photographer** Anan Naruphantawat

태국 수완나품 공항 근처 마을에 위치한 2층짜리 단독주택은 256㎡ 대지의 3분의 2에 불과한 150㎡의 크기로 자리 잡고 있다. 부부와 두 아이, 할머니, 총 다섯 식구가 살기에는 터무니없이 작았다. 이 곳은 작은 곡선의 삼각형 대지 형태일 뿐만 아니라 도로 양쪽과 연결되는 모퉁이에 위치해 있다. 공간을 최대한 활용하기 위해 대지의 곡선과 평행한 만(灣) 모양으로 확장했다. 3층은 필요한 모든 기능을 결합할 수 있도록 만들고, 안쪽으로 들어간 공간은 휴식 연못이 있는 야외 작은 조경으로 활용할 수 있다.

전체 건물 높이에 영향을 미치는 보의 깊이를 줄이기 위해 철골구조를 적용하고 기둥의 크기를 줄임으로써 실내 공간을 넓히는 데 도움을 주었다. 동선을 최소화하기 위해 두 개의 계단을 각기 다른 모서리에 배치했다. 계단 사이에 채광창을 설치한 이중 공간은 사생활 보호가 부족한 수직적 빈 공간을 만드는 대신 아래층으로 자연 채광이 들어 오도록 했다.

프라이버시가 필요한 모퉁이에 위치한 집인 만큼 외피 커버가 필요하다. 수직 알루미늄 격자와 타공 금속을 함께 사용하여 기존 주택으로 이어지는 모든 만을 덮어 두 건물이 함께 조화를 이루도록 하였다. 증축 건물은 기존 건물보다 층고가 높기 때문에 두 건물을 연결하기 위해 경사진 파사드 형태가 필요했다. 하지만 기존 주택은 단순한 평면 형태만으로도 충분하기 때문에 이 부분에 대해서는 비스듬한 입면을 만들 필요가 없었다. 유지보수 문제를 해결하기 위해 상층부 파사드에는 분말 코팅 알루미늄을 사용했는데, 이는 가볍고 고층부에서 보수나 재도장이 어렵기 때문에 유지보수가 필요 없다. 그러나 태국에서는 가격이 저렴하면서도 유지보수가 용이한 금속을 집 벽에 사용했다.

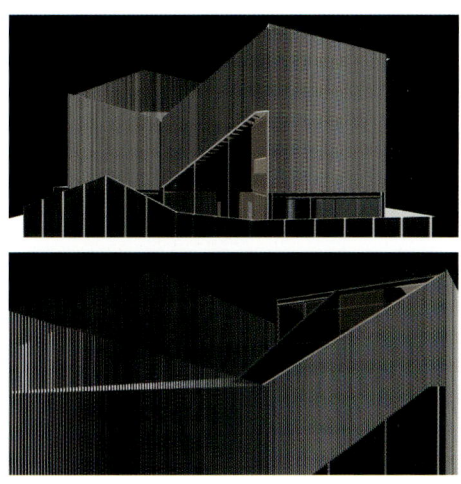

ISOMETRIC

↑ Corner view ↵ Back view

↑ Front view

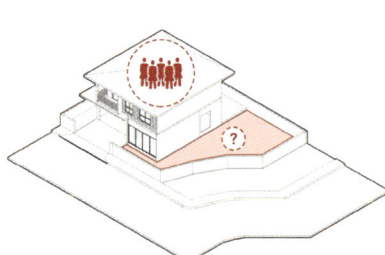

01 A COVE LAND
A two-story single-detached house. It was too small for a family with five members ; thus, it is turned into another extension house.

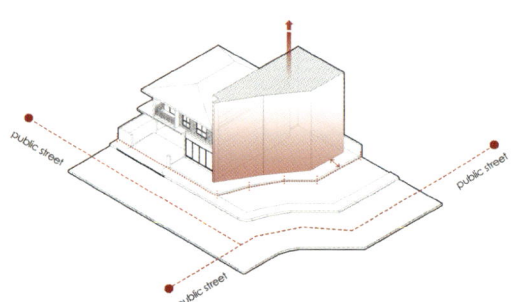

02 MAXIMIZE A USED SPACE
A small curve-triangular land shape and located at the corner which connects to both sides of public street , in order to use the space to its full potential.

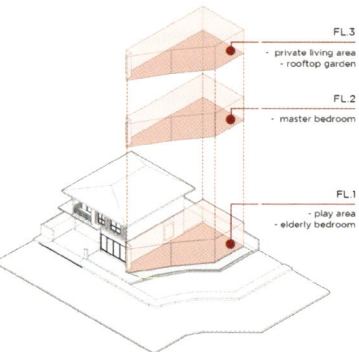

03 OLD 2 AND NEW 3
Three-floor is created to combine all functions needed, which are; an activity space connects to an existing house and outdoor terrace.

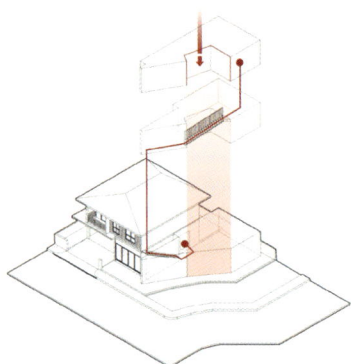

04 SEESAW STAIRS / SKYLIGHT
Two staircases are placed in different corner which minimize circulation. Double space in between the stairs with skylight allows natural sunlight through downstairs.

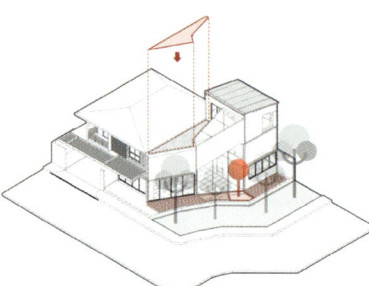

05 GREEN REPLACEMENT
Since an existing garden is replaced by the building, it is elevated to the top floor instead. to remain green area space of the house.

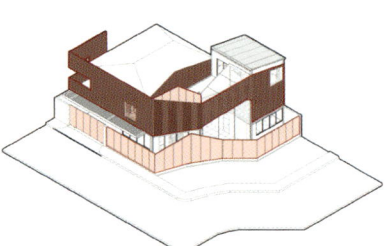

06 FACADE COVER
As the house located at the corner, façade cover is required. Vertical aluminum trellis together with perforated metal are used to cover all.

DIAGRAM

← Night view → Exterior view

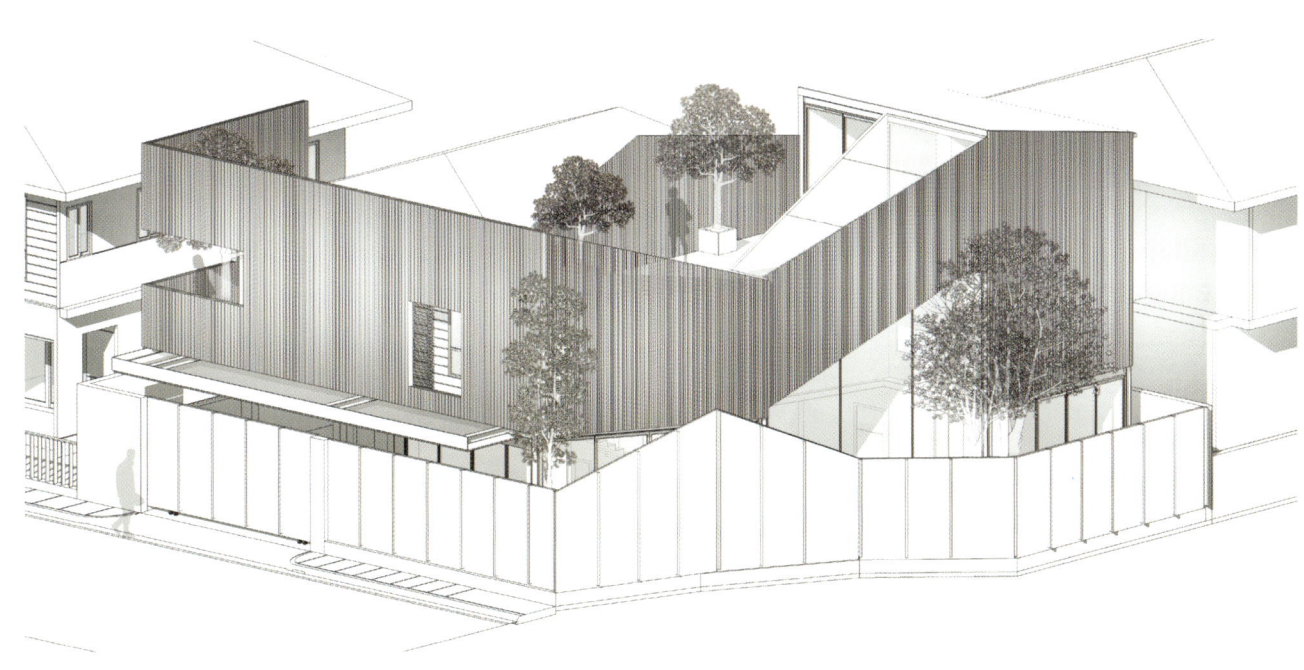

ISOMETRIC

 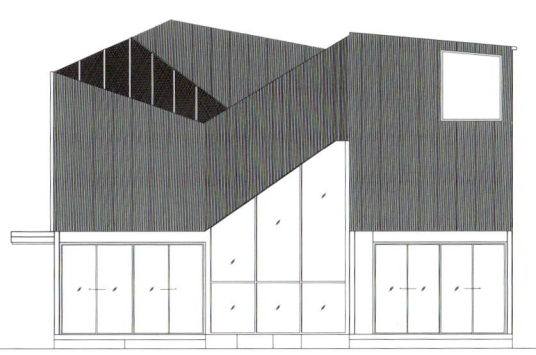

ELEVATION

↑ Play area

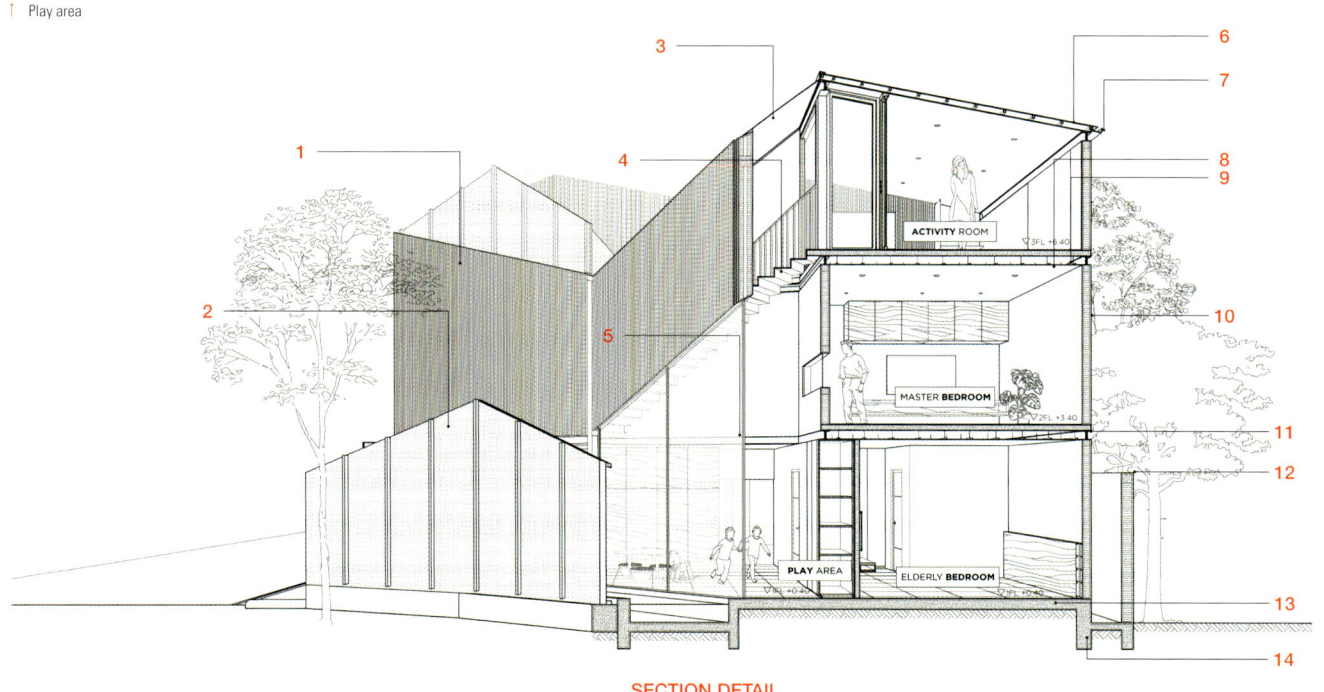

SECTION DETAIL

↑ Play area & Side garden

1 25*50mm VERTICAL WHITE RECTANGULAR ALUMINIUM TRELLIS
2 WHITE PERFORATED METAL ON STEEL STUDS
3 LAMINATED GLASS SKYLIGHT
4 FOLDED STEEL PLATE STAIRS WITH MATTE URETHANE POLISHED ASH WOOD FINISHING
5 TEMPERED GLASS WINDOW
6 SNAP LOCK METAL SHEEY ROOF WITH PU FOAM INSULATION
7 STAINLESS STEEL ROOF GUTTER
8 T9mm GYPSUM BOARD CEILING WITH WHITE PLASTERED AND PAINTED
9 ALUMINIUM C-LINE STUDS
10 LIGHTWEIGHT CONCRETE WALL
11 STEEL I-BEAM
12 EXISTING CONCRETE BLOCK WALL
13 REINFORCED CONCRETE SLAB WITH GRANITO FLOOR TILE FINISHING
14 REINFORCED CONCRETE BEAM
15 WOOD PLASTIC COMPOSITE DECK ON STEEL JOINT
16 STEEL STRUCTURE STAIRS WITH MATTE URETHANE POLISHED ASH WOOD FINISHING

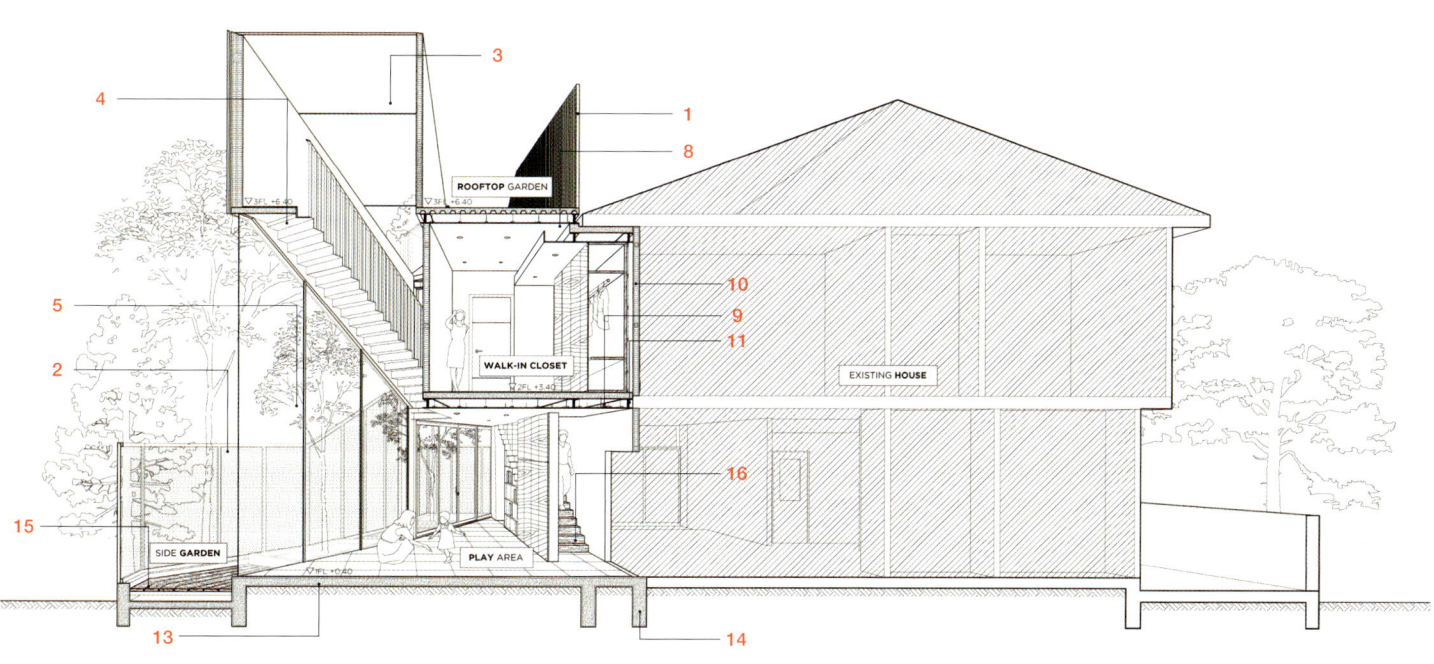

SECTION DETAIL

← Side garden   → Staircase   ↙ Bedroom

↑ Living room

1 EXISTING HOUSE
2 CONNECTING DOOR
3 PLAY AREA
4 STAIRCASE
5 ELDERLY BEDROOM
6 BATHROOM
7 TERRACE
8 POND
9 READING AREA
10 WALK-IN CLOSET
11 MASTER BEDROOM
12 LIVING ROOM & FITNESS
13 PANTRY
14 ROOFTOP GARDEN

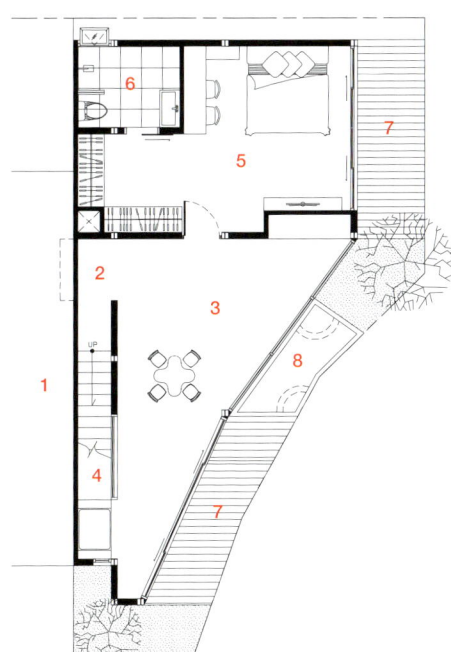

1ST FLOOR PLAN

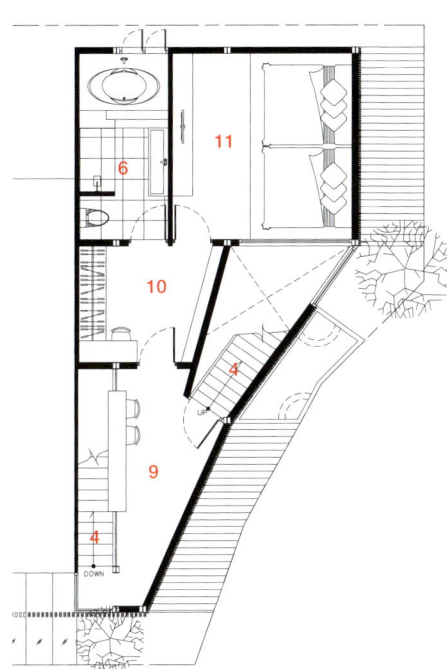

2ND FLOOR PLAN

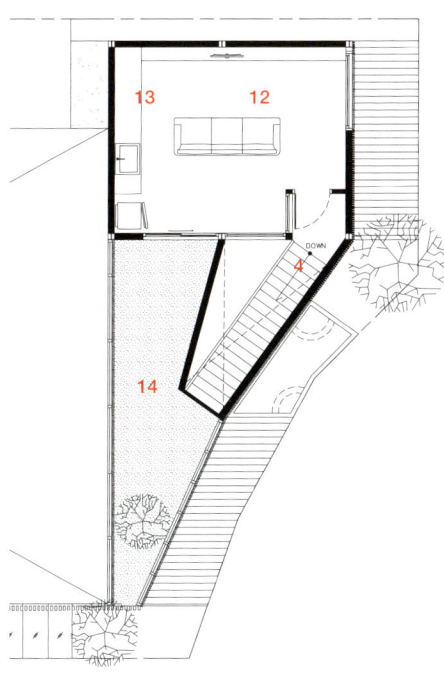

3RD FLOOR PLAN

THIS EXPERIMENTAL PROJECT IS LOCATED IN the first belt of low-density neighborhoods surrounding downtown Denver at only two miles of the city center. It provides centrally located, low-cost housing for individuals or couples, while integrating within the morphology of the suburban environment. The site consists of a 50 foot wide parcel divided into two equal lots. According to local zoning codes, we could built a main house and an 'accessory dwelling unit' (ADU) on each lot. By organizing each front house into 3 studios, each with its own bathroom and kitchenette, and a large communal living space, we managed to have eight units in total: six studios in the front houses, and two split-level 'artist' studios in the ADU's towards the alley.

The project acknowledges how larger single-family residences in well-located neighborhood are frequently shared by roommates and friends, and was designed to cater to those needs. Integrating shared kitchen and living room, laundry areas, a powder room, and paved outdoor areas, the project stages a subtle balance between the need for privacy and the possibility of social interaction.

The project is built with a very limited budget (about 200 usd/sqft) using economical construction materials and standard solutions. The pitched roof volumes are clad with a standing seam metal roof (in a standard blue color) and a board and batten façade with different vertical intervals and tones of blue, to make the child-like house-shapes recognizable elements. Even within the restricted budget and buildable volume, the architecture has a generous and spacious feel. The front houses have double-height entrance areas, and the communal kitchen and living areas have large glazed surfaces to interact with each other. The upper-floor studios take advantage of the pitched roof volume creating high-ceiling living areas and a sleeping mezzanine accessible by a ladder. The split-level ADU's received a roll-up garage door creating a workshop-like atmosphere and a direct connection with the outdoors during the warm summer months.

Location Denver, Colorado, USA **Use** Hosing **Surface** 332.63m² **Local architect & contractor** Joe Dooling (DDB) **Collaborators** Ruy Berumen, Emiliano Rode, Tessa Watson **Developer** Continuum Partners **Photographer** Onnis Luque

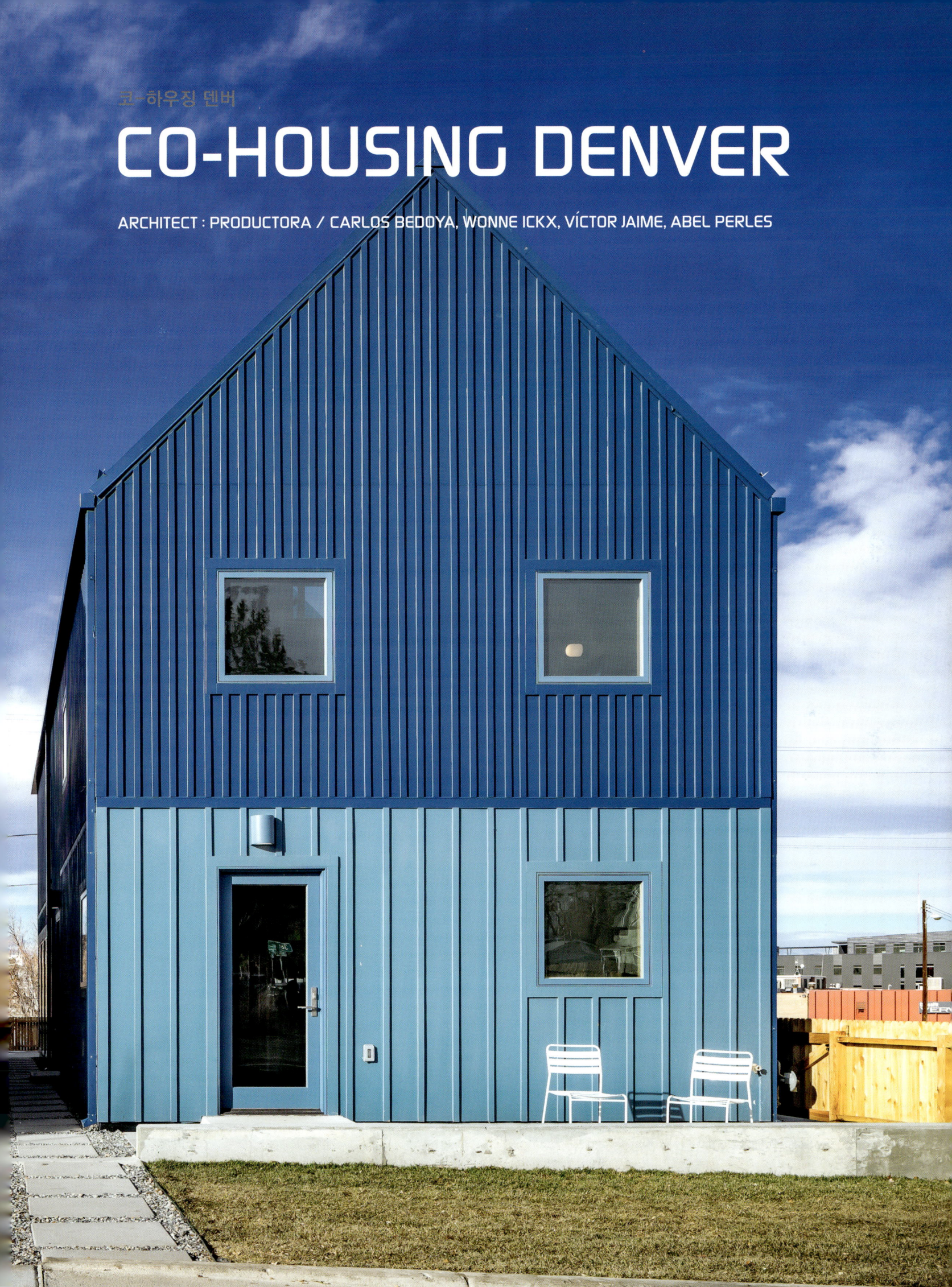

CO-HOUSING DENVER

ARCHITECT : PRODUCTORA / CARLOS BEDOYA, WONNE ICKX, VÍCTOR JAIME, ABEL PERLES

↑ Exterior & street view   ↵ Top view

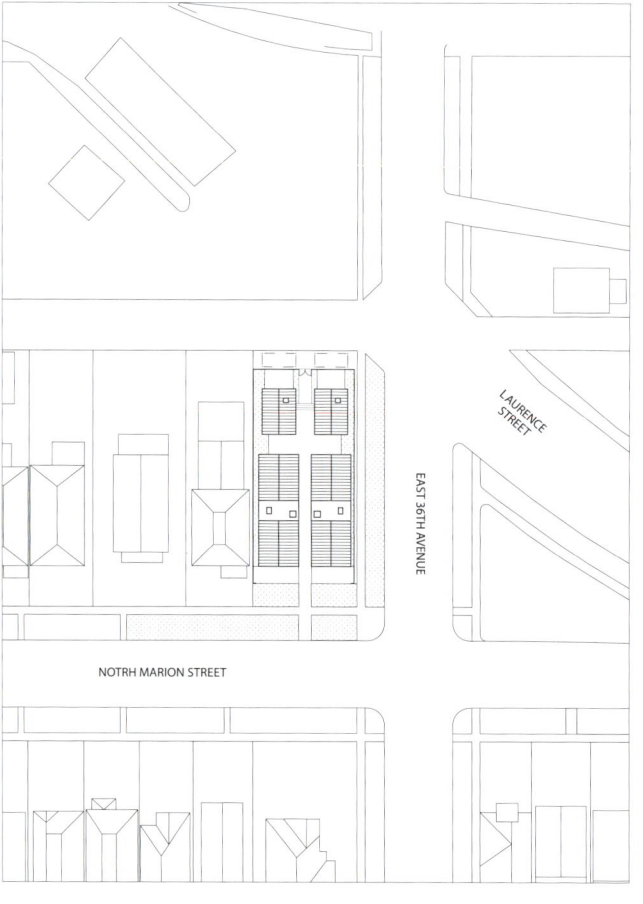

SITE PLAN

↑ Panorama

EAST ELEVATION

WEST ELEVATION

이 실험 프로젝트는 도심에서 불과 2마일 떨어진 덴버 시내를 둘러싼 첫 번째 저밀도 지역 벨트에 있다. 개인 또는 커플을 위한 저렴한 비용의 중심가 주거 공간을 제공하면서, 동시에 교외 환경의 형태학적 특성에 통합된다. 부지는 폭 50피트의 토지가 두 개의 동등한 필지로 나뉜다. 지역 건축 규정에 따르면, 각 필지에 주택과 '부속 주거 유닛'(ADU)를 건설할 수 있다. 각 전면 주택을 자체 욕실과 간이 주방이 있는 3개의 스튜디오로 구성하고 넓은 공동 거실 공간을 마련함으로써, 전체 여덟 개의 유닛을 관리하게 되었다. 전면 주택에 여섯 개의 스튜디오와 골목 쪽의 ADU에 두 개의 분할 레벨 '아티스트' 스튜디오가 포함되어 있다. 이 프로젝트는 좋은 위치 동네에 있는 큰 단독 주택들이 자주 룸메이트나 친구들과 공유되는 점을 인지하고, 그러한 요구를 충족하도록 설계되었다. 공용 주방과 거실, 세탁 구역, 화장실, 포장된 야외 공간을 통합하여, 개인의 사생활 보호 필요성과 사회적 교류 가능성 사이의 미묘한 균형을 프로젝트에서 연출하였다.

이 프로젝트는 제한된 예산(약 200달러/평방피트)을 사용하여 경제적인 건축 자재와 표준 해법으로 구축되었다. 경사진 지붕 구조물은 표준 파란색의 스탠딩 솔기 금속 지붕으로 마감되었고, 다양한 수직 간격과 푸른 색조의 판재와 루버 외관을 사용하여, 아이들의 그림 같은 집 모양을 인지할 수 있는 요소로 만들었다. 제한된 예산과 건설할 수 있는 부피에도 불구하고, 건축물은 넉넉하고 넓은 느낌을 제공한다. 전면에 위치한 주택들은 두 배 높이의 입구 공간을 갖추고 있으며, 공용 주방 및 거실은 상호 작용할 수 있도록 큰 유리 면적을 활용하였다. 상층부 스튜디오는 경사진 지붕 구조를 이용하여 높은 천장의 생활 공간을 만들어 내고, 사다리를 통해 접근할 수 있는 침실 중이층을 갖췄다. 분할 수준의 ADU는 롤업 차고 문을 설치하여 작업장 같은 분위기와 따뜻한 여름 동안에는 실외와의 직접 연결이 가능하다.

← West, main gate   → Public road

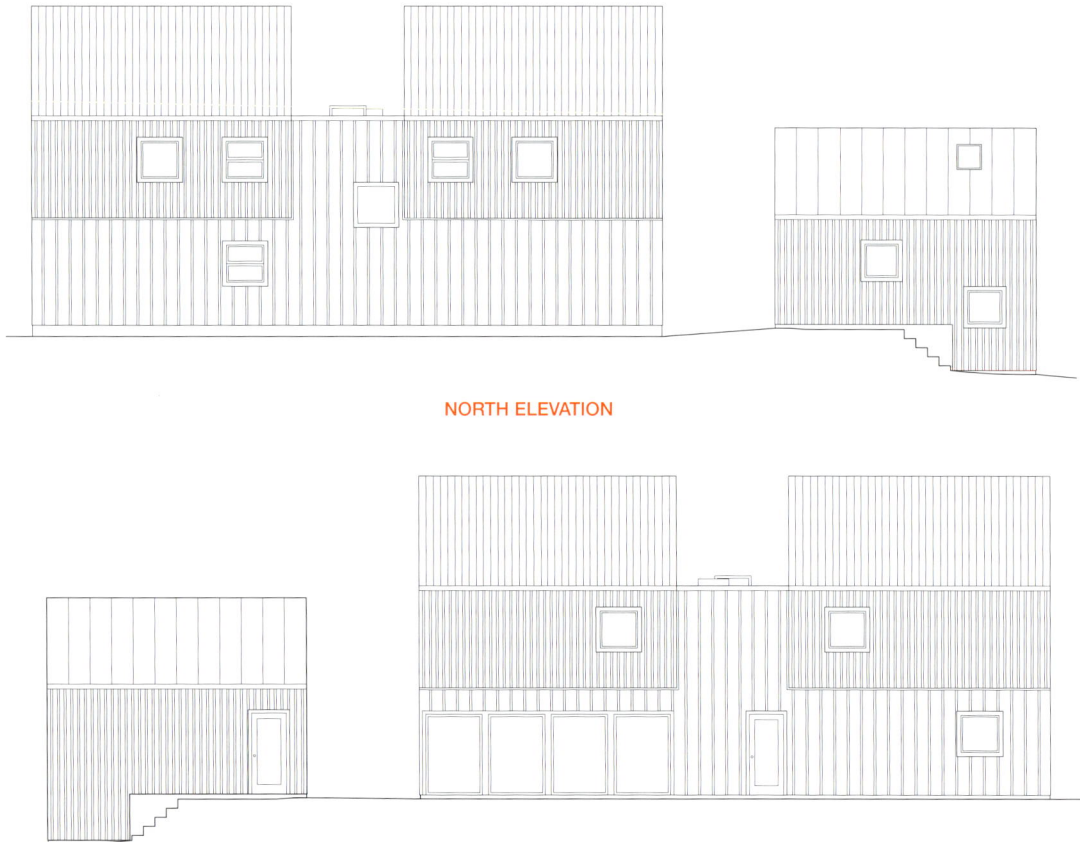

NORTH ELEVATION

SOUTH ELEVATION

← Roll-up garage door   → Blue color board & batten facade

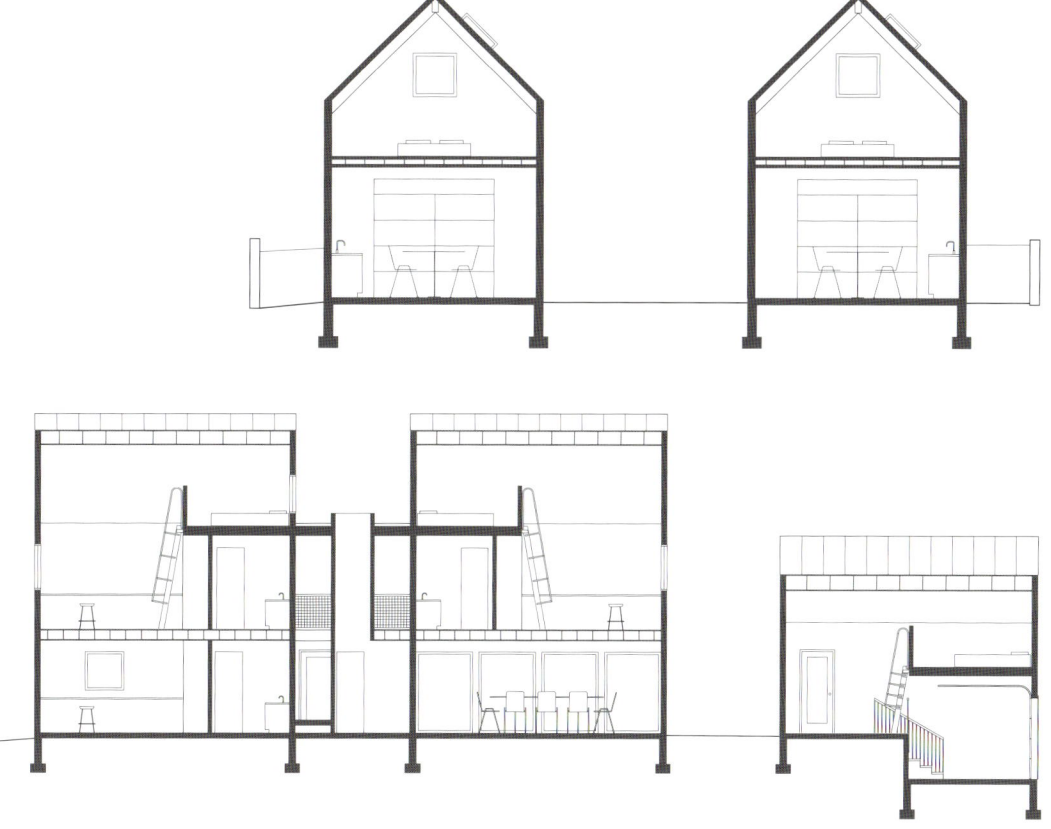

SECTION

↑ Living areas & large glass

1 SHARED KITCHEN	3 TOILET	5 PRIVATE SUITE	7 LAUNDRY
2 SHARED LIVING & KITCHEN	4 ACCESS	6 ADU	8 BEDROOM

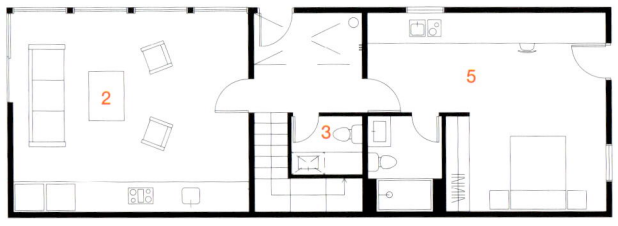

GROUND FLOOR PLAN

← Communal kitchen   → High-ceiling living areas & a sleeping mezzanine

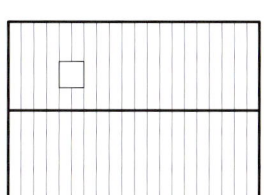

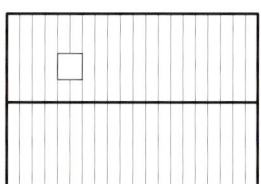
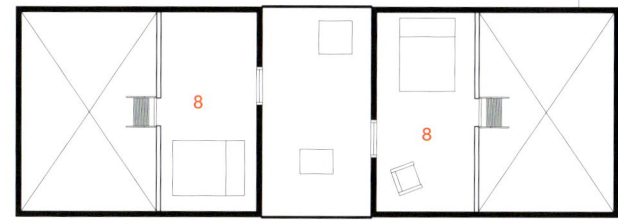

3RD FLOOR PLAN

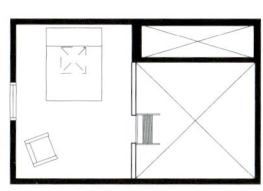
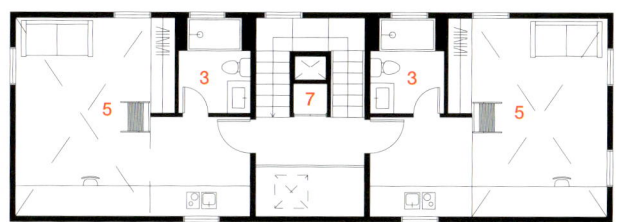

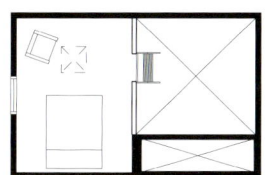

2ND FLOOR PLAN

아파트 에스
APARTMENT S

ARCHITECT : TSUKAGOSHI MIYASHITA SEKKEI / TOMOYUKI TSUKAGOSHI, JUMPEI MIYASHITA

THIS IS AN APARTMENT COMPLEX THAT HAS 3 FLATS. After covid-19, we feel the anxiety of sharing space with others. The apartment consists of 1 basement and 2 stories and has the car parking on the site. The lower flats have the worse light and wind conditions. Therefore, it is tried to reduce the inequalities by bending the ceiling and planning a window besides the bent ceiling.

The basement is the studio for the owner. It has a bend in the ceiling and a wide high window along the western street. This creates a comfortable studio with a lot of indirect sunlight. And the ceiling in the east is also bent to create the entrance of the basement. The ground and first floors are apartment units and the positions of the windows in these units are decided by considering the privacy of the residents. The ceiling, which contributes to the positions of the windows and the privacy of the residents, is bent in the northwest and southwest corners of the ground floor and the tall high window in the corners faces each crossing point outside. Especially the window in the southwest corner sticks out and becomes a bay window to expand the interior space.

The floor plans of the above residential floors were designed subject to the existence of the bent floor slabs. The existence of those floor slabs creates unique spaces on the above floor. On the long sloped floor slab on the ground floor, there has been put a countertop and it has become a long multi-use fixed furniture piece. However, on the first floor, the bent floor slab in the northwest corner has become the back of the sofa bench and in the southwest corner, the bent floor slab has become a large windowsill. What is more, the bent ceiling of the entrance of the basement makes a difference in the angle of the outside stairs. And that creates terraced steps with a low slope in the upper part. The bend ceiling slab of the lower flat appears on the above floor and affects the way to live for the resident there.

Location Tokyo, Japan **Use** Apartment **Site area** 84.51m² **Built area** 45.81m² **Gross area** 121.04m² **Project manager** Tomoyuki Tsukagoshi **Design team** Tomoyuki Tsukagoshi, Jumpei Miyashita **Contractor** SANRYO **Photographer** Jumpei Suzuki

이 프로젝트는 3개의 플랫이 있는 아파트 단지이다. 코로나19 이후로, 우리는 다른 사람들과 공간을 공유하는 데 대한 불안감을 느끼고 있다. 아파트는 지하 1층과 지상 2층으로 구성되었으며, 부지 내에 차량 주차 공간이 있다. 하층 플랫은 조명과 통풍 여건이 더 좋지 않았다. 따라서, 천장을 굽히고 굽힌 천장 옆에 창문을 계획함으로써 불평등을 줄이려고 노력하였다.

지하실은 소유주를 위한 스튜디오로, 서쪽 거리를 따라 천장에 굴곡이 있고 넓고 높은 창문이 있다. 이에 따라 많은 간접 햇빛이 들어오는 편안한 스튜디오가 만들어졌다. 그리고 동쪽의 천장도 지하 입구를 만들기 위해 굽혀져 있다. 지상층과 첫 번째 층은 아파트 유닛이며, 이 유닛들의 창문 위치는 거주민의 사생활을 고려하여 결정됐다. 창문의 위치와 거주민의 사생활에 기여하는 천장은 지상층의 북서쪽과 남서쪽 모서리에 굽혀져 있고, 모서리의 높은 창문은 바깥쪽의 각 교차점을 향하고 있다. 특히 남서쪽 모서리의 창문은 돌출되어 베이 창문이 되며 내부 공간을 확장했다.

위 주거층의 평면도는 굴곡층 슬래브의 유무를 고려하여 설계되었다. 이러한 바닥 슬래브의 존재는 위층에 독특한 공간을 만들어 냈다. 지상층에 있는 긴 경사진 바닥 슬래브 위에는 카운터탑을 두어 긴 다용도 고정 가구로 활용되었다. 하지만 1층에는 북서쪽 모서리에 구부러진 바닥 슬래브가 소파 벤치의 등받이가 되고 남서쪽 모서리는 구부러진 바닥 슬래브가 큰 창틀이 되었다. 더욱이, 지하실 입구의 굽은 천장은 바깥 계단의 각도에 차이를 만들어 상부에 완만한 경사의 테라스 계단을 생성하였다. 아래층 플랫의 구부러진 천장 슬래브는 위층에 나타나며 거주자의 생활 방식에도 영향을 준다.

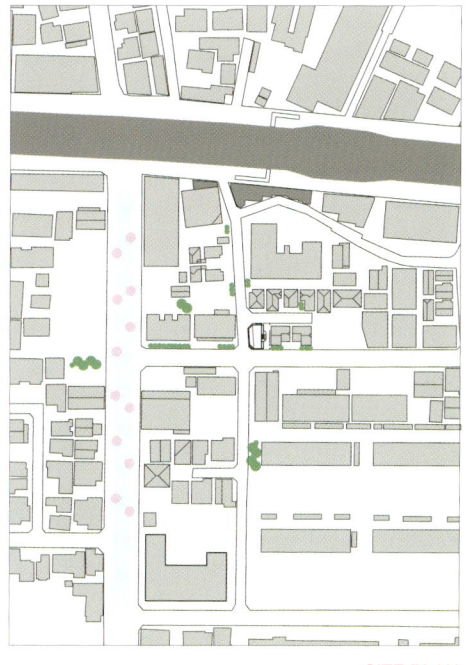

SITE PLAN

← Exterior view

← South, high bay window → North-west, tall high window

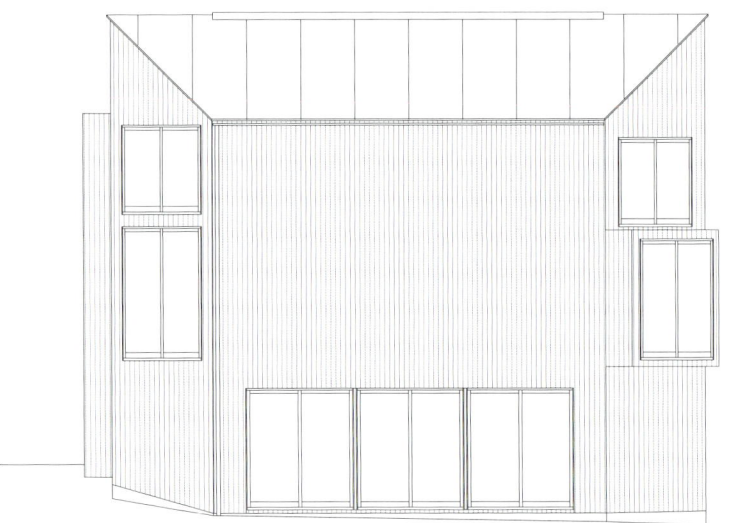

WEST ELEVATION

SOUTH ELEVATION

↑ West exterior, wide & tall high window

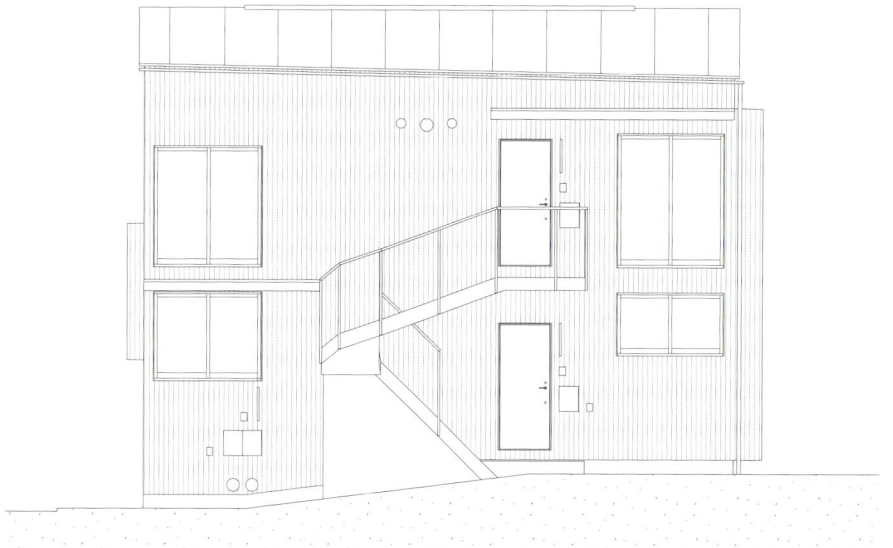
EAST ELEVATION

← Enterance & outside stairwell   → Window detail   ↙ Entrance & stair

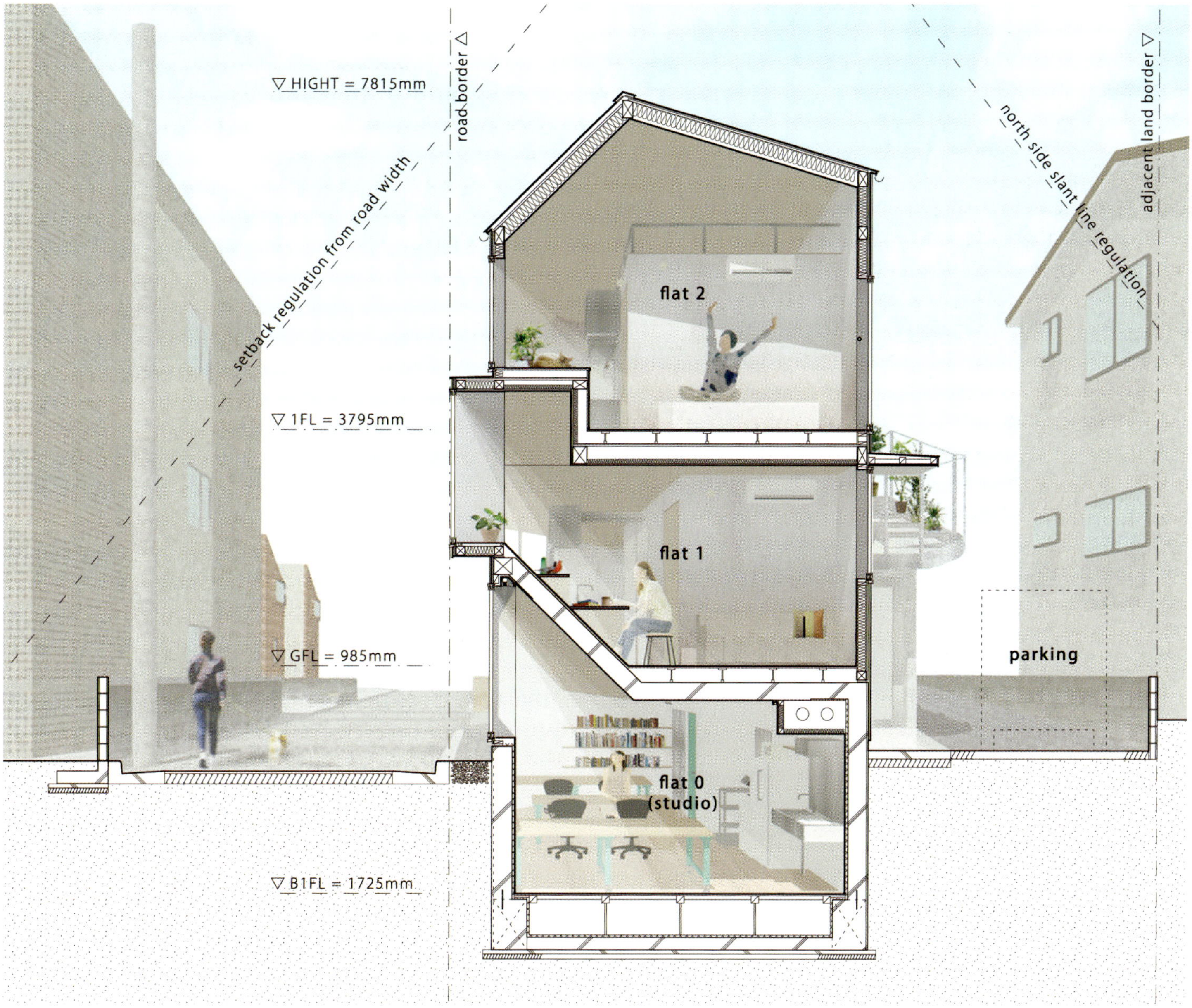

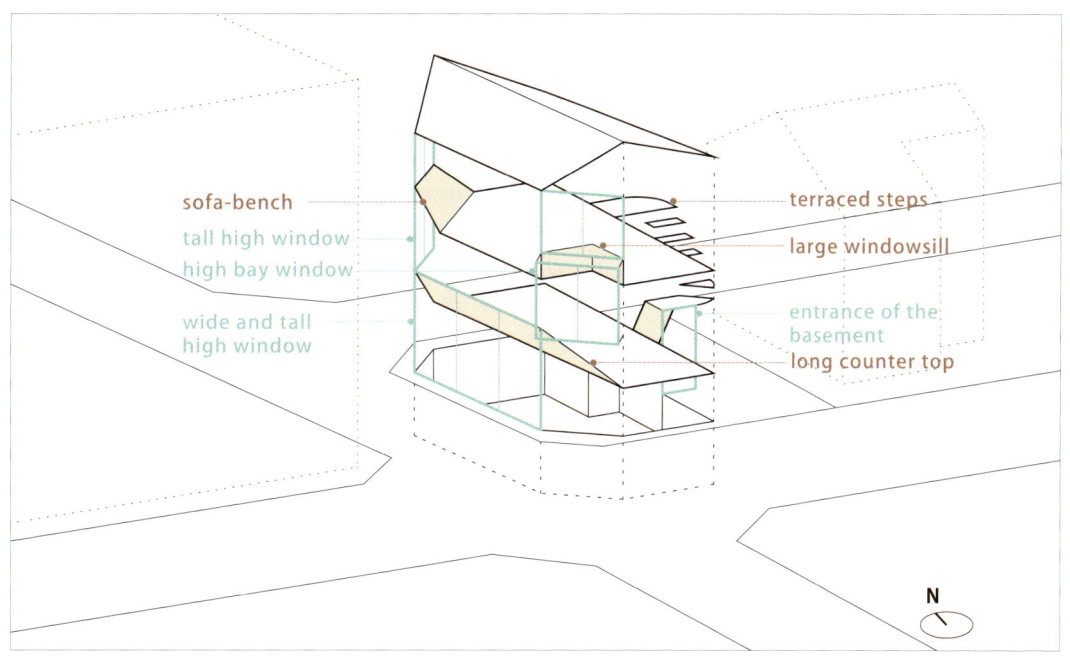

SECTION DETAIL

DIAGRAM

↑ Basement floor studio

INTERIOR ELEVATION

↑ 1st floor long multi-use fixed furniture   ↙ 1st floor kitchen furniture system   ↓ 1st floor long multi-use fixed furniture   ↘ 1st floor long multi-use fixed furniture

↑ 2nd floor room   ↙ 2nd floor southwest corner, the bent floor slab has become a large windowsill

1	PARKING	6	BATHROOM
2	ENTRANCE	7	ROOM
3	TOILET	8	STUDIO
4	KITCHEN	9	STORAGE
5	WASHBASIN		

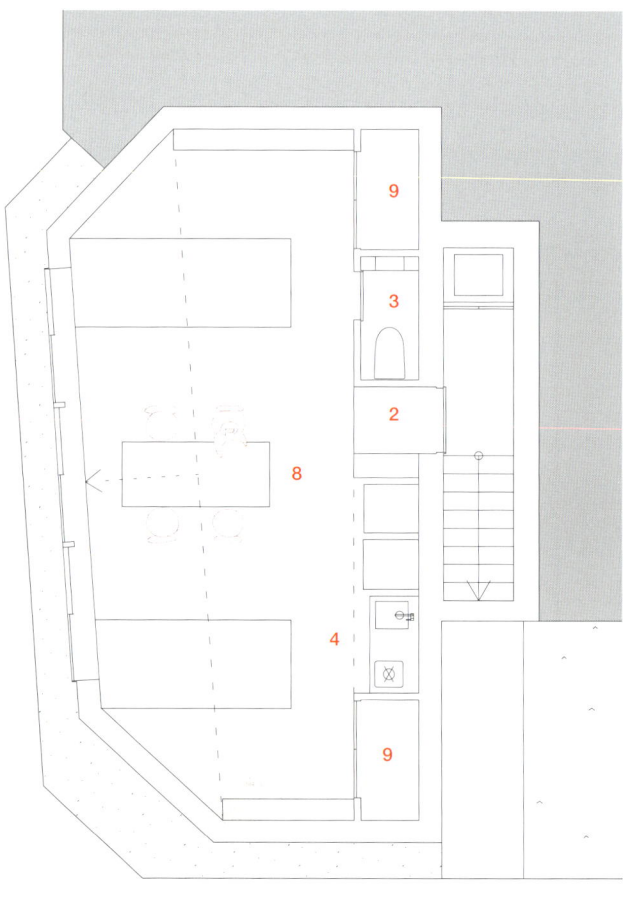

BASEMENT FLOOR PLAN

← 1st floor, northwest corner has become the back of the sofa bench → 1st floor, northwest corner the sofa bench

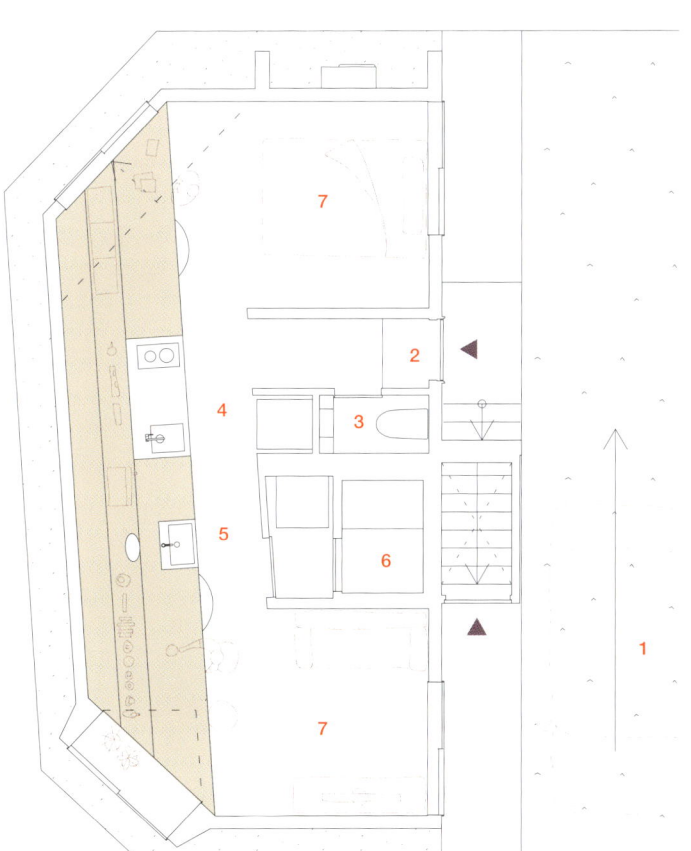

GROUND FLOOR PLAN

1ST FLOOR PLAN

히엔 하우스
THE HIÊN HOUSE

ARCHITECT : WINHOUSE ARCHITECTURE

THE HIEN IS LOCATED IN A NEWLY-DEVELOPED SUBURBAN AREA of Danang city, in which you can enjoy a panoramic view of the picturesque Han river. This home is a new place where a 3-generation family enjoys a cozy life together. Woodworking is their traditional profession, which has been passed down through generations, and it serves as the main inspiration for choosing wood as the primary material for this project. This decision not only provides comfort and nostalgic but also maximizes cost savings, as the homeowners themselves participate in the woodworking process for the construction.

The rural village is deeply rooted in the childhood memories of the family's generations. The design team does not want to disrupt their familiar landscape, lifestyle, and customs when transitioning to the urban environment, the house's structure incorporates typical elements of "rural architecture". In Vietnamese, "Hiên" means a veranda, which is a covered space with a roof, serving as a buffer zone between the interior and exterior of the house, it is not only an accent but also a traditional architectural feature that suits the tropical climate of Central Vietnam. Some functions are arranged similar to the layout of their countryside home, as the team believes that habituation is the most critical element.

The design centers around the relationship between humans and nature. Green spaces are seamlessly incorporated into the living areas, establishing a seamless connection between nature and daily activities, This enables a comfortable space for beneficial organisms while deterring pests and harmful creatures by planting fragrant herbs that repel them effectively.

The HIÊN is not just a house; it is a symbol of love and passion for the family's traditional craft, a place that strengthens family ties. By blending modern and traditional elements in harmony with nature, the team aims to leave a valuable mark on the cultural essence of the homeland amidst the ever-developing urban context.

Location Hoa Quy, Ngu Hanh Son, Da Nang, Vietnam **Use** Housing **Site area** 350m² **Completion** 2023 **Project manager** Hoang Phan Quoc Huy, Thai Huu Hai **Manufacturers** INAX, SAT, Samsung **Structural Engineers** Bim City **Photographer** Quang Dam

히엔 하우스는 그림 같은 강줄기의 전경을 감상할 수 있는 다낭 시의 새로 개발된 교외 지역에 위치한다. 이 집은 3대가 함께 아늑한 삶을 즐기는 새로운 장소이다. 목공은 대대로 전해져 내려오는 그들의 전통적인 직업이며, 이 프로젝트에서 주 재료로 목재를 선택하게 된 주된 영감의 역할을 하였다. 이 결정은 편안함과 향수를 제공할 뿐만 아니라, 건축주가 시공에 있어 직접 목공 작업에 참여함으로써 비용 절감을 극대화하였다.

시골 마을은 가족 세대의 어린 시절 추억에 깊이 뿌리를 두고 있다. 설계 팀은 도시 환경으로 전환할 때 익숙한 풍경, 라이프 스타일 및 관습을 방해하고 싶지 않으며 집의 구조는 "농촌 건축"의 전형적인 요소를 통합한다. '히엔(Hien)'은 베트남어로 베란다를 의미한다. 베란다는 지붕으로 덮인 공간으로 집 내외부 사이의 완충지대 역할을 하며 악센트뿐만 아니라 중부 베트남의 열대 기후에 맞는 전통적인 건축 특징이다. 설계 팀은 습관화가 가장 중요한 요소라고 믿기 때문에 일부 기능은 시골 집의 레이아웃과 유사하게 배열하였다.

디자인은 인간과 자연 간의 관계를 중심으로 한다. 녹지 공간은 생활공간에 매끄럽게 접목시켜 자연과 일상생활을 원활하게 연결해준다. 이를 효과적으로 밀어내는 허브를 심어 해충과 유해생물을 억제하는 동시에 유익한 생물이 살기 좋은 공간을 확보할 수 있다.

히엔 하우스는 단순히 집이 아니라 가족의 전통 공예에 대한 사랑과 열정의 상징이자, 가족 유대를 강화하는 장소가 된다. 현대적이고 전통적인 요소를 자연과 조화롭게 융합함으로써 끊임없이 발전하는 도시 맥락 속에서 고국의 문화적 본질에 가치 있는 흔적을 남기고자 한다.

← Front facade → Terrace

↑ Bird's eyes view ↵ Night view

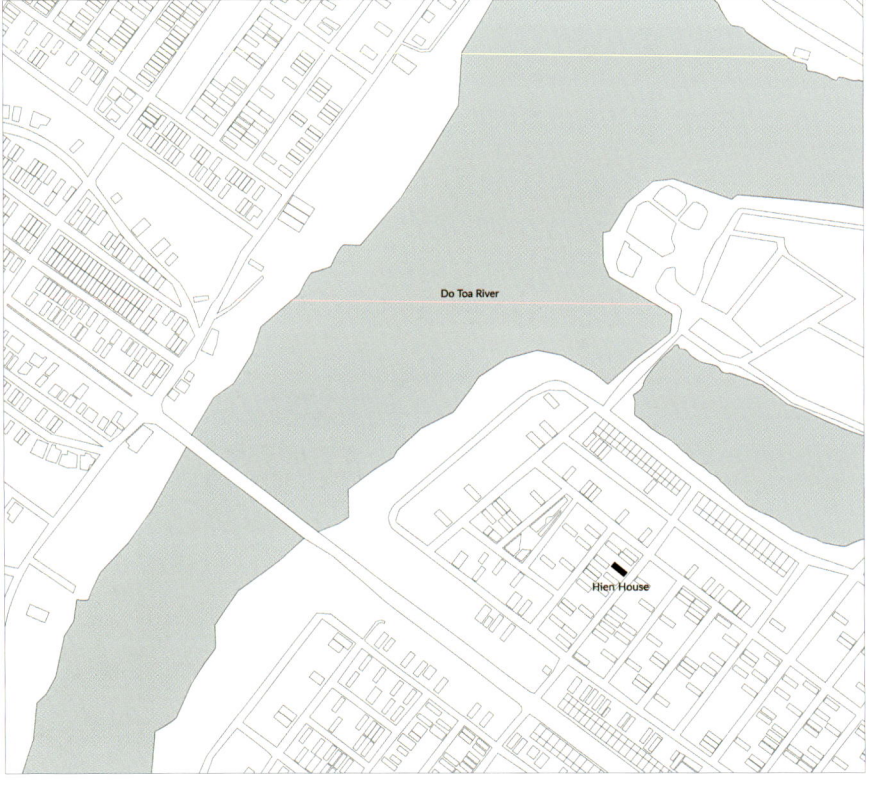

SITE PLAN

← Side view → Rear view

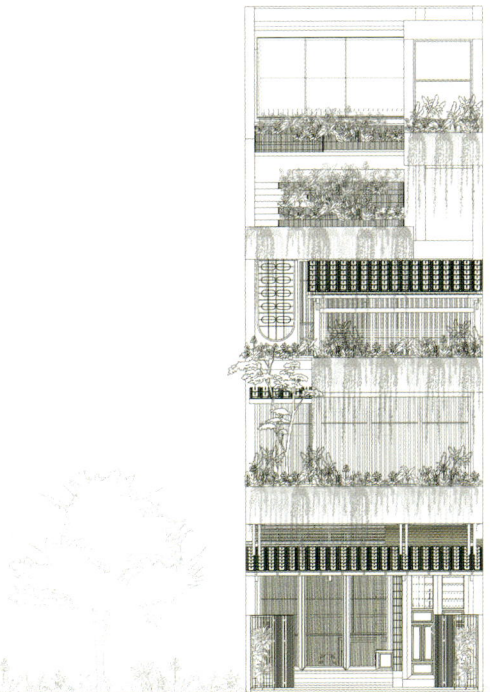 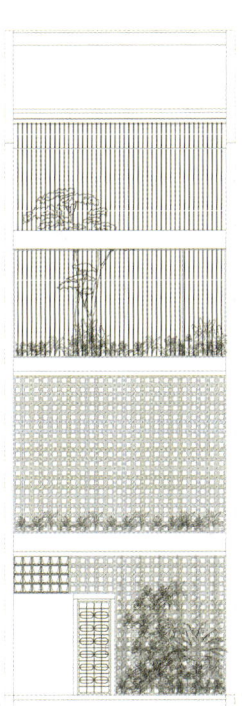

FRONT ELEVATION **BACK ELEVATION**

↑ Main entrance

CROSS SECTIONAL PERSPECTIVE

← Wood working → Yard & Traditional roof

1 STAINLESS STEEL M10 BOLT
2 A CONNECTING COMPONENT WITH A THICKNESS OF 10mm
3 STAINLESS STEEL BELT BUCKLE
4 WOODEN RIDGE BEAM FOR THE SUSPENDED ROOF SYSTEM
5 STAINLESS STEEL M8 BOLT
6 WOODEN RIDGE BEAM FOR THE SUSPENDED ROOF SYSTEM HAS DIMENSIONS OF 120*60mm
7 WOODEN BEAMS WITH DIMENSIONS OF 100*60mm SPANNING ACROSS THE SUSPENSION MEMBERS OF THE ROOF SYSTEM
8 VIGLACERA HA LONG TILES
9 ROOFING SHINGLES 30*15mm
10 ROOFING TILES WITH A SIZE OF 60*30mm
11 WOODEN RAFTERS WITH DIMENSIONS OF 80*40mm
12 WOODEN DIAGONAL ROOF BEAMS WITH DIMENSION OF 100*600mm

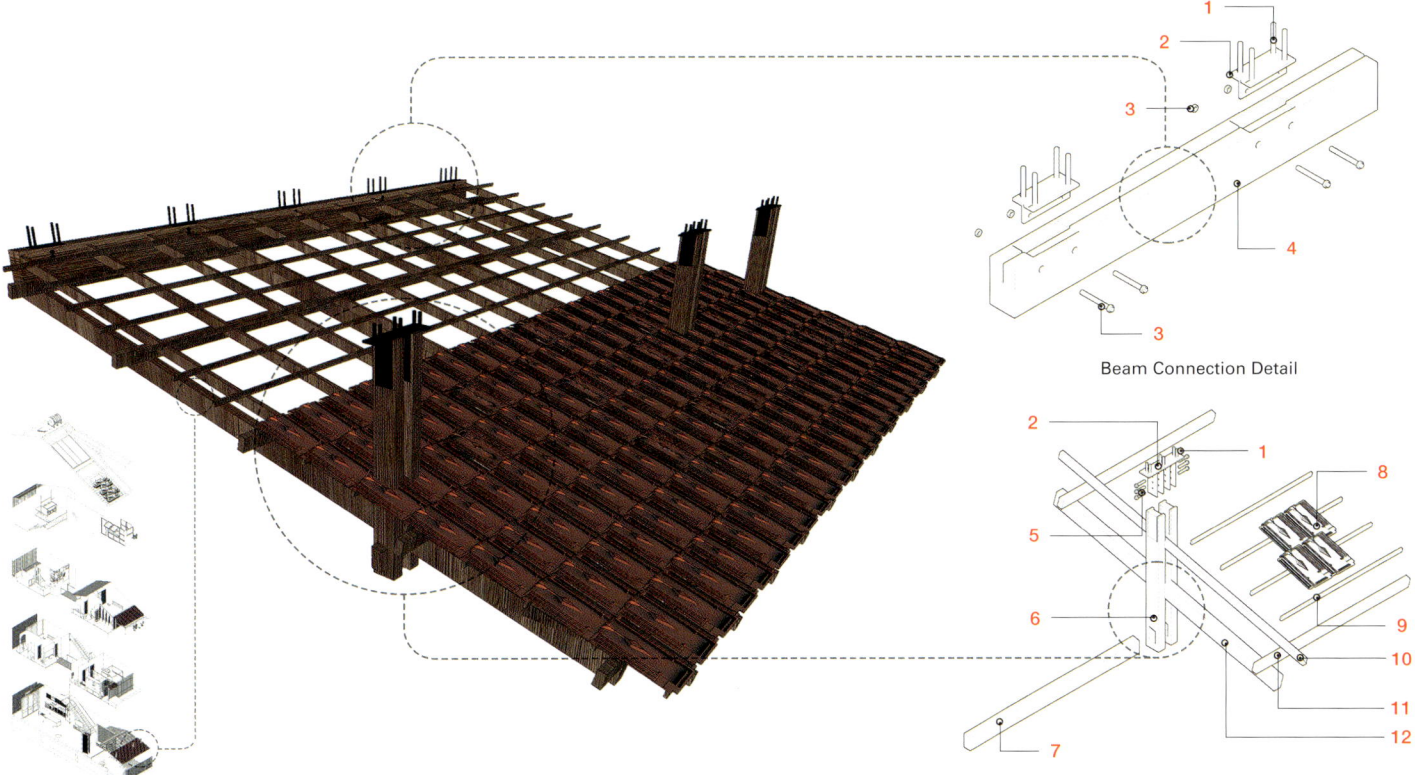

Beam Connection Detail

The Detailed connection of the suspended roof framing system involves

THE SUSPENDED ROOF SYSTEM
Instead of using traditional support methods such as wall beams or large columns, we have opted for an inverted suspended roof system. This roof extends beyond the reinforced concrete floor above and provides shelter from rain and sun for the outdoor parking area. Half of the roof is covered with tiles on the outer side, while the inner side remains open to allow for ventilation, increased natural light, and air circulation.

BEAM CONNECTION DETAILS
The connection between wood and the concrete ceiling can be achieved using steel components, while the remaining wood connections can utilize traditional joinery methods.

STRUCTURAL COMPOSITION

← Patio → Terrace

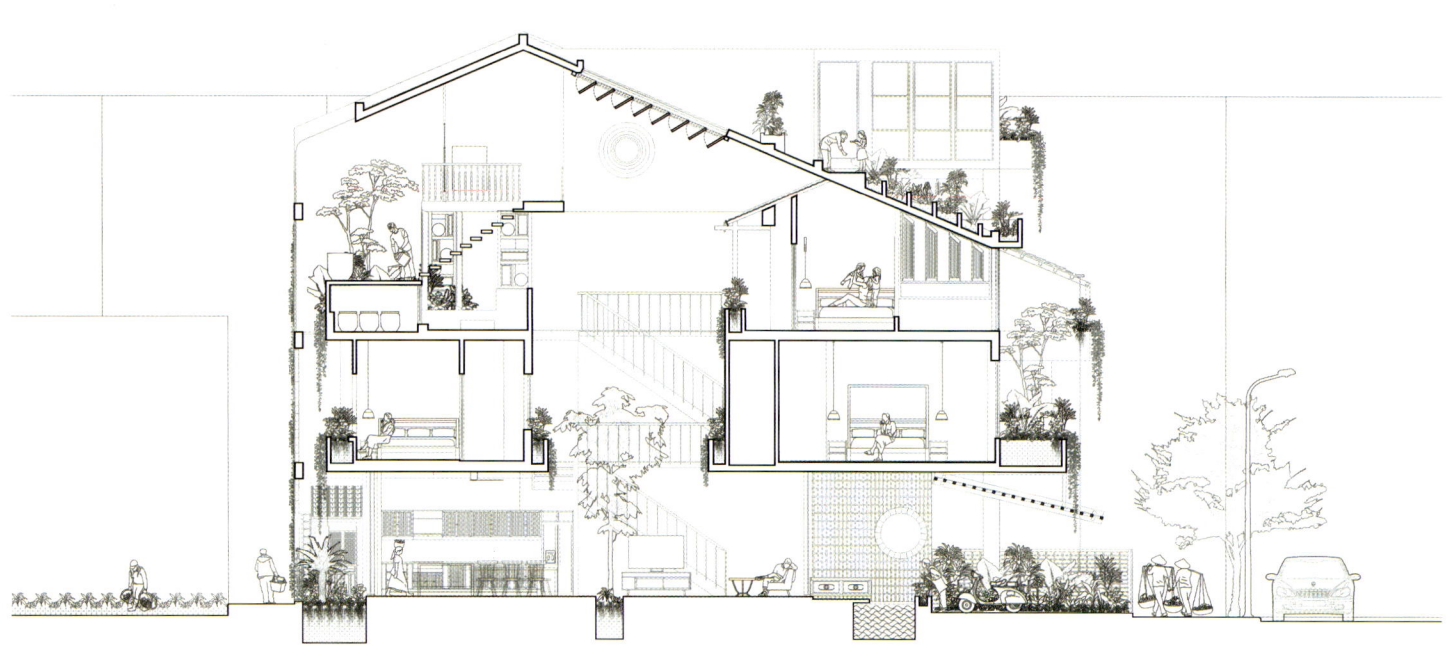

SECTION

← Bamboo shade   → Interior view

1 THE REINFORCED CONCRETE FOUNDATION COMPONENTS - STAIRCASE BASE
2 EINFORCED CONCRETE PODIUM
3 SOLID WOOD PODIUM
4 WALL-MOUNTED WOODEN HANDRAIL
5 STAINLESS STEEL M8 BOLT
6 WOODEN POST SECTION OF 120*120mm
7 8mm THICK STEEL PLATE, WELDED TO CREATE A SUPPORT FRAME FOR SUN SHADING
8 WOODEN HANDRAIL
9 30*5mm FLAT BAR STEEL
10 WOODEN PLANK
11 Ø12 REFERS TO A STEEL BAR WITH A DIAMETER OF 12mm, CB 300
12 WOOD VENEER ON THE OUTSIDE OF AN I-BEAM STEEL
13 A STEEL CONNECTING COMPONENT WITH A THICKNESS OF 10mm
14 WOODEN BEAMS CONNECTED TO SUSPENDED REINFORCED CONCRETE BEAMS FOR SHADE STRUCTURE
15 STAINLESS STEEL M10 BOLT
16 WOODEN STEP
17 TIMBER STEP 980*270*30mm
18 STEEL FRAME SYSTEM CONNECTING 2 WOODEN POSTS
19 WOODEN POST SECTION OF 120*120mm
20 10mm STEEL PLATE EMBEDED PARTIALLY IN CONCRETE, CONNECTING REINFORCED CONCRETE FOUNDATION WITH WOODEN POST

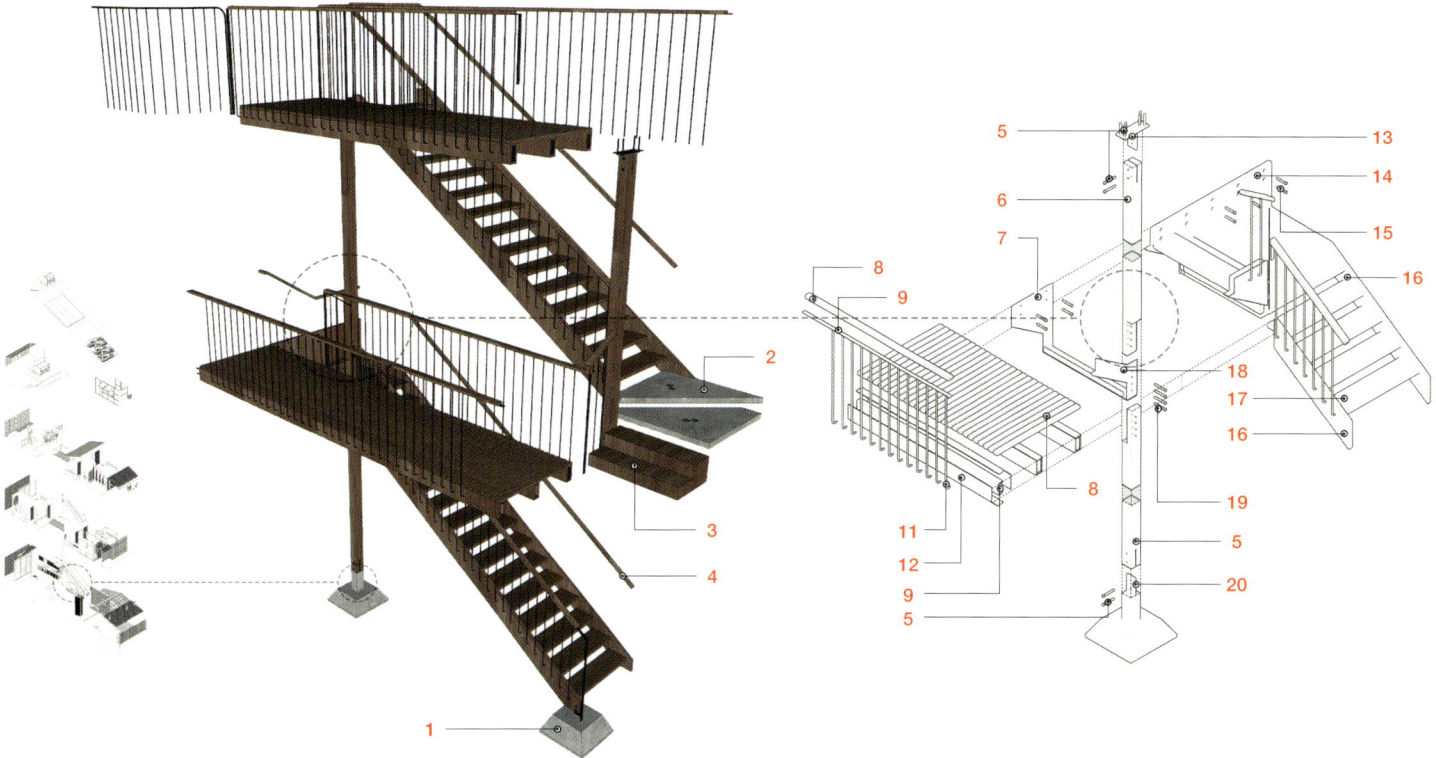

WOODEN LADDER SYSTEM
The wooden staircase system is supported by wooden posts, securely connected to the floor, and reinforced with steel components and expansion bolts to the concrete beams. The wooden components are sourced from reclaimed wood, obtained through collecting timber from old structures with local regional significance.
This approach aims to promote resource conservation by utilizing existing available resources.

LINKING CONSTRUCTIONS
The details of this staircase are quite intricate, with the connections between wooden, steel, and concrete components forming a robust vertical transportation system.

STRUCTURAL COMPOSITION

↑ Kitchen & Dining room

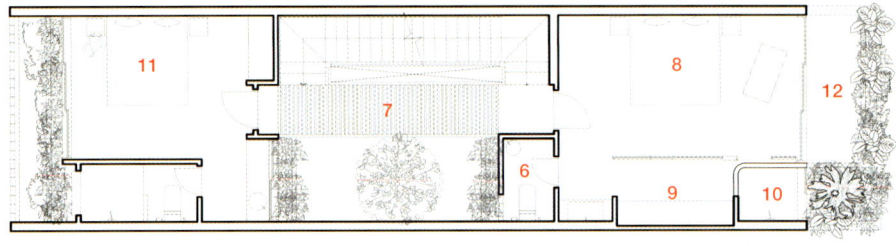

1 YARD	6 TOILET	11 BEDROOM	16 CREATIVE ROOM
2 ENTRANCE	7 HALLWAY	12 TERRACE	17 VEGETABLE GARDEN
3 RELAXATION SPACE	8 MASTER BEDROOM	13 OUTDOOR BATHTUB	18 ALTAR
4 POND	9 WALK-IN CLOSET	14 LAUNDRY	19 SKYLIGHT
5 DINING & LIVING ROOM	10 BATHROOM	15 STORAGE	20 WATER TANK

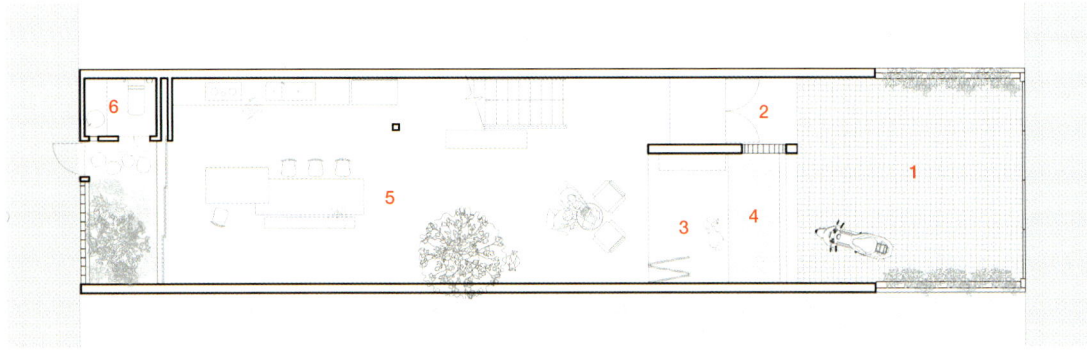

2ND FLOOR PLAN

1ST FLOOR PLAN

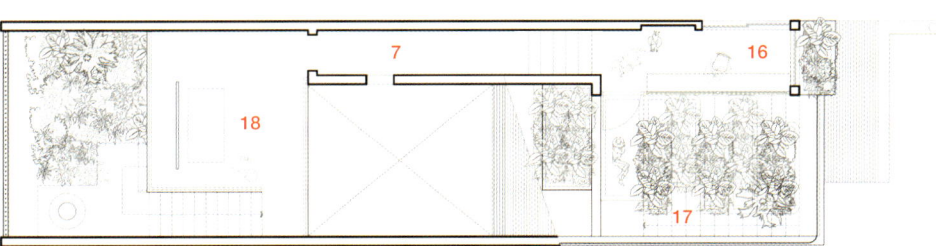

ROOF PLAN

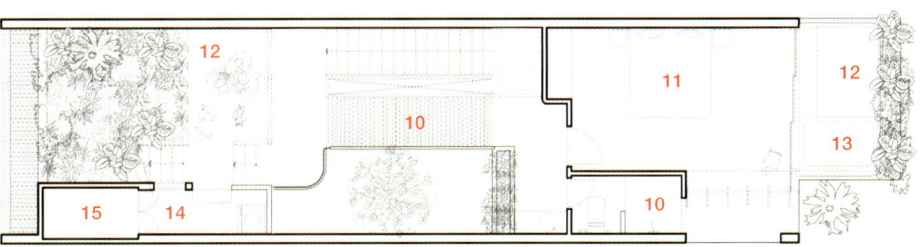

4TH FLOOR PLAN

3RD FLOOR PLAN

PROFILE

↘ page 004, 072

Andrew Maynard, Mark Austin | Austin Maynard Architects

Andrew Maynard is a Tasmanian now living and working in Fitzroy, Australia. Andrew Maynard Architects was established in 2002 after Andrew won the Asia Pacific Design Award's grand prize for his mobile work station, THE DESIGN POD. Distinctive, exciting and radical, Andrew is also a founding board member of Nightingale Housing - a non-for-profit organisation creating ethical, socially sustainable and cost effec-tive housing, whilst also revolutionising the developer-dominated housing market.

Mark Austin is also a staunchly proud Tasmanian. He joined Andrew Maynard Architects in 2007 and become a director of the practice in 2009. In 2016 the name was changed to Austin Maynard Architects to officially recognise his contribution. Mark has enjoyed a diverse career since graduating from the University of Tasmania in 1993 and Melbourne University in 1996. He spent six years working in London, as Pro-duction Designer for the English National Opera and for a number of commercial firms specialising in a variety. Throughout his career Mark continues to show a incredible balance between design tal-ent and rigorous, pragmatic, technical and organisation skills.

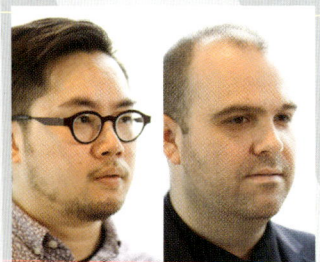

↘ page 018

Jarrett Boor, Daniel Yao | Paperfarm Inc

Paperfarm Inc was established in 2014 by designers Jarrett Boor and Daniel Yao as a design intensive architecture practice headquartered in NYC. With a diverse range of work in varied scale, Paperfarm seeks to explore design solutions to complex program and site constraints in urban and non-urban settings.

With studios in North Stonington, Connecticut and Taipei, Taiwan, we have completed single family residences in America and Asia and work in mixed-use, multi-family and commercial design.

↘ page 028

Florian Busch | Florian Busch Architects

Florian Busch Architects (FBA) is an office practicing architecture and urbanism. Based in Tokyo, FBA draws on a global network of the industry's top experts, ensuring that a wealth of knowledge and innovation drives every project to solutions beyond the imagined. FBA's works comprise a wide spectrum of structures and buildings at all scales. FBA's research team focuses on socio-cultural analysis as well as research and development of new materials and building technologies.

↘ page 040

Jenchieh Hung, Kulthida Songkittipakdee | HAS design and research
HAS design and research was founded by Jenchieh Hung and Kulthida Songkittipakdee, and they explore Asia's architectural language through a "design + research" parallel approach. Hung And Songkittipakdee (HAS) work encompasses cultural buildings, religious architecture, installation art, exhibition design and experimental projects. They are also actively involved in academia as visiting professors and design critics in architecture at Tongji University, Chulalongkorn University, and King Mongkut's University of Technology Thonburi.

↘ page 052, 244

Setthakarn Yangderm, Parpis Leelaniramol | TOUCH Architect
TOUCH Architect, founded in 2014 by the architects, Setthakarn Yangderm and Parpis Leelaniramol, which is a design studio based in Bangkok. It was changed from mini studio into a company with a great chance of an improvement. Founders are collaborating with the team in design, construction, and management.

"TOUCH" comes from our expertise and services which cover all project stages. It focuses on basic human needs together with cultural, environmental concern and context experiment, with an aesthetic of sustainability architecture. An ordinary structure and simple material are used to form an architecture, as well as to create functional and practical dwellings.

Full scope of design services includes all various types of project phases, from clients' disposition, project research and concept, design development, construction drawing and permission, construction inspection, as well as, technical supervision.

↘ page 062

Nuno Campos | TRAÇO ALTERNATIVO arquitectos associados, lda.
TRAÇO ALTERNATIVO arquitectos associados, lda. is a company based in Porto, founded by partners Nuno Campos and Pedro Cardoso in 2004. It provides services in the areas of Architecture, Rehabilitation, Consulting and Construction. It has a multidisciplinary work team with extensive professional experience.

Using BIM technology since its foundation, the company has always been at the forefront of technology in the use of complex and demanding software for parametric and three-dimensional representation, thus enhancing creativity, accuracy and project control.

↘ **page 086**

Pramoth Kitkanasiri | Spacy Company Limted

Spacy architecture is an architectural and interior design company based in Bangkok, Thailand. Our practice focuses on space design facilitating functional and visual connectivity that create dynamism and blur boundary between interior and exterior.

↘ **page 098**

Miaojie Ted Zhang, Anton Schneider, Andri Luescher | Found Projects + Schneider & Luescher

Found Projects is a Los Angeles-based architecture and design studio founded by Ted Zhang and Veronica Smith. Our practice evolved from building an archive of projects that influenced us, to sharing each other's work, to eventually collaborating on built work. Our approach is formally and tectonically driven and influenced by our respective backgrounds.

Schneider & Luescher is a mono-disciplinary architecture studio based in Los Angeles, California. Operating at the intersection of Jesse & Clarence.

↘ **page 108**

Marcos Hagerman | HGR arquitectos

HGR ARQUITECTOS is an architecture and design studio founded in 2009. Located in Mexico City, HGR seeks to actively participate in the development of the city through a design that integrates the social, economic and geographical aspects of each project.

The office, despite having been established recently, has developed numerous projects, each with specific characteristics. In each of these projects HGR ARCHITECTS has worked from the initial concept and the design of furniture and image, to the execution of the work, always seeking contributions from experts and specialists in each branch and integrating the client's vision, budget and times to the design process.

At HGR, we consider the importance of details, as well as the application of various materials to innovate construction processes as a fundamental part of our methodology.

↘ page 116

Eugene Seow | CDG Architects
CDG Architects is an architectural practice based in Singapore devoted to crafting spaces and creating places. We specialise in landed residential projects of all scales, including interior design and architecture, and aim to imbue our works with the fundamentals of good design and quality of space. Our wider portfolio includes alternative energy solution projects, hospitality and mixed-use buildings, competition-winning institutional and commercial buildings, and master-planning projects, all while working closely with allied professionals and trades.

↘ page 124

Langjin Zhu | Archipoetry Studio
Archipoetry Studio was established in 2014 by Langjin Zhu. "Archipoetry" is a way of thinking and an attitude of space creation that focuses on the combination of thinking on essential matter and poetic expression. Archipoetry Studio has expand itself into an architectural studio that integrates lifestyle brands. Gradually explicit the working strategy of "living, construction and researching". The lifestyle brand is the foundation, construction is the core and research work is the mission of the studio.

↘ page 136

Luis Carlos Aguilar Gonzalez | HERYCO
HERYCO, Architectural design firm established in Mexico by Luis Carlos Aguilar, handcrafting every project in a BIM environment.

↘ page 144

Sangkyong Jeong, Inkeun Ryoo, Doran Kim | YOAP architects Ltd.
YOAP architects is a group planning, designing, marketing; architecture/ interior/ items. We aim not to limit architectural ideas to buildings but to extend it to diverse design areas. Seeking for sustainable enjoyments in design works, we also hope to share it. As its name, YOAP ('nearby' in Korean) will hold an image of friendly neighbor.

↘ page 154

Nguyen Van Thien | TAA Design

TAA Design is an architectural firm founded by Nguyen Van Thien in 2017. He graduated from the University of Architecture in Ho Chi Minh City in 2012. Then he worked as an architect and an artist who specialized in calligraphy painting.

Nguyen Van Thien focuses on creating architectural projects closely connected to nature and the surrounding environment. His most famous project is The Red Roof which has received numerous international awards for its significance.

↘ page 162

Hao-Chun Hung | Üroborus_StudioLab

Born in Taipei. Hao-Chun Hung founded Üroborus_StudioLab in 2017. He was working for Heatherwick Studio as senior designer and site architect. Üroborus_StudioLab is a young design studio based Taipei, Taiwan. The studio is combining Urban, Architecture, Space, Terroir, Vision, Interaction, Craftsmanship and local technique for design propose. He graduated Master of Architectural Design, The Bartlett School of Architecture, University College London, UK and he has teach Lecturer of Department of Art & Design, Yuan Ze University.

↘ page 174

Chalermchai Asayote | Sata Na Architect

Sata Na Architect is a company in Bangkok suburb. The core idea of our designs is "Architectural design in consideration of location, livelihood, time, and way of life." We focus on interior and exterior designs. We have worked as consultant ideas of design, analysis lifestyles of users, design concept of building, designing and site visiting. Our mission is to capture happiness in each steps of architect design form sketching owner's idea to owner's move in building.

↘ page 182

Thanawin Pattanawong | V2in Architects

V2in Architects Limited is a Bangkok-based small-scale architecture firm specializing in upscale residential design. Thanawin Pattanawong, its chief architect and founder, aspires to create simple and truthful architecture that represents the essence of functionality, locality, and user behavior. V2in Architects approaches sophisticated requirements with minimalist aesthetics and attentive choices of masses, forms, and spaces.

↘ page 194

Doan Thanh Ha | H&P Architects

Doan Thanh Ha (1980) graduated from Hanoi Architectural University (HAU) in 2002, in 2007 he holds a masters degree in HAU. He set up and have been operating H&P Architects since 2009. His works have been receiving high appraisals in Vietnam and overseas and prizes including UIA Vassilis Sgoutas Prize, UIA Turgut Cansever Award, Barbara Cappochin Prize, Architizer A+ Award, World Architecture News Award (WAN), International Architecture Award (IAA), Architectural Review House Award (AR), ARCASIA Awards for Architecture (AAA), AZ Award, National Architectural Award,...

↘ page 204

Chandrakant Kanthigavi | 4site architects

As a founder and Principal architect, CK always insisted not just on great design at the drawing board but the flawless execution of the project. He believes in an open work culture, communication and deep collaboration within the firm and outside firm ensures delighted clientele at the end of every project. With 18 years of experience he believes that As an Architect, one needs to design an environment that enhances life, rather than just being a shelter. All of the project have a vision of creating spaces that are more than shelters and they shape the clients behaviour and life style.
He is recognised as Young Designer 2016 by Indian Architect and Builder, Firm's work published many National and international architectural journals such as A+D magazine, Archdaily, Archello, Inhabitat, RTF and Buildofy.

↘ page 212

Vinh Phuc Ta | ROOM+ Design & Build

ROOM+ Design & Build is an award-winning multi-disciplined firm with intensive expertise in architectural, interior, landscape design as well as construction and project management. The practice has worked on many projects in Vietnam and Australia and achieved prestigious design awards, including Architecture Masterprize; Dezeen Awards; and World Architecture Community Awards. ROOM+ Design & Build always sensitively responds to the briefs and the contexts in order to propose creative design and construction solutions which enrich spatial experience, optimize functionality and effectiveness.

↘ page 220

Anna Jach | Studio Anna Jach GmbH

Studio Anna Jach GmbH is a multidisciplinary architecture, design and art practice established in Zurich, Switzerland in 2012.

Founded on over twenty-five years of work experience on projects of various scales, from highrise buildings, bridges, and cultural buildings to house renovations and interior design, our studio understands architecture as a process of shaping the cultural, social and physical environment. Due to the dramatic climate changes in the previous years, we put our strong focus on ecology, sustainability, re-use of materials and reduction of construction waste. Anna Jach holds a Certificate of Advanced Studies in Regenerative Materials from ETH Zurich, where she was exploring tools and methods to manage projects using earth, bio-based materials like straw, hemp, lime and rammed earth with efficiency and creativity to contribute to the change of the architectural mindset. Each of our projects has a strong identity and authenticity and it's based on an active dialog between the client and the architect. Our current large-scale project is a cultural center in Warsaw, Poland, including a concert and event hall, learning rooms, rooftop cinema, participative garden and basketball court.

↘ page 228

Eric Yu, Chen-Tien Chu | Atelier SUPERB + Chen-Tien Chu

Atelier SUPERB is an award-winning design studio based in Taichung, Taiwan. Our expertise transcends the confines of traditional disciplines, embracing architecture, interior spaces, landscapes, urban planning, artistic expression, exhibitions, and branding. We firmly believe that the essence of professionalism lies in the seamless integration of design thinking and effective communication. The journey from conception to creation is not merely a process - it's a dynamic movement, an active catalyst propelling us towards progress and shaping the contours of the future.

Chen-Tien Chu Architect was Founded in 1995, the design philosophy of Chen-Tien Chu Architect is centered on exploring the intrinsic complexion of space, drawing inspiration from the nature and its history. We strive to deliver in a context-driven approach, harnessing light from the surroundings to infuse vitality into our designs, resulting in spaces that exude a rich poetic expression.

↘ page 236

Bill Tsui | Yu2e

Yu2e is an architecture and engineering practice providing design and consultation services for buildings and places in Los Angeles since 2007. Every project is an opportunity to discover the inherent potential of the interrelated complexities amongst site, clients, occupants, community, builders, budgets, schedules, and our own curiosity. Yu2e has engaged projects in a wide spectrum of roles including lead design consultant, a sub-consultant on project design teams (engineer, exec. arch, code consult, etc.), owner-builder, and in-house architect for developers. Essentially, we are the ultimate team player either leading or supporting projects, typically all the way to delivery. Yu2e maximizes engagement to facilitate a broader influence for design and invention. Yu2e navigates the often irrational and nonlinear process of feasibility, design, entitlement, development and construction to reveal rich opportunities for project innovation, quality and execu

↘ page 252

Carlos Bedoya, Wonne Ickx, Víctor Jaime, Abel Perles | PRODUCTORA
PRODUCTORA is working on projects in Mexico and abroad, ranging from residential projects to public buildings. PRODUCTORA has been awarded by the Architectural League of New York with the Young Architects Forum (2007) and the Emerging Voices (2013). In 2016 they received the Mies Crown Hall Americas Prize for emerging architects for the 'Pavilion on the Zocalo' at the IIT in Chicago. The Teopanzolco Cultural Center, received the American Architecture Prize in Cultural Architecture (2017), Oscar Niemeyer Award 2018, and the First International Prize at the Biennial of Architecture of Quito 2018.
PRODUCTORA has been actively involved in teaching both in local Universities such as the Universidad Iberoamericana, Centro de Diseño, TEC de Monterrey and Universidad La Salle in Mexico City as well as abroad.

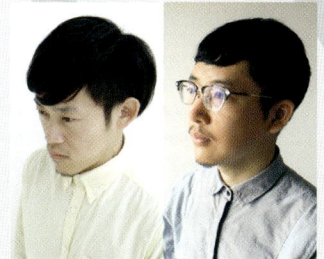
↘ page 260

Tomoyuki TSUKAGOSHI, Jumpei MIYASHITA | Tsukagoshi Miyashita Sekkei
Tsukagoshi Miyashita Sekkei is the architecturel studio based in Tokyo. The studio is established in 2015 by Tomoyuki TSUKAGOSHI and Jumpei MIYASHITA.

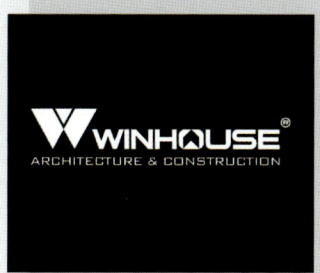
↘ page 270

WINHOUSE Architecture & Construction
WINHOUSE Architecture & Construction is a total design - construction and construction consulting firm. With over 10 years of experience. We specialize in hospital buildings, resorts, hotels, apartments, offices, housing.

Publisher | Heungchae Jung
Editorial Dept. | Joonyong Jung, Eunjae Ma
Design Dept. | A&C design team

Print in Korea
ISBN | 978-89-7212-006-3
Price | USD 68 (68,000won)
Registration No. 2004-000166

© A&C Publishing
9F, 15, Teheran-ro 22-gil, Gangnam-gu, Seoul, Korea
T: +82-2-538-7333
www.ancbook.com

Copyright A&C Publishing and may not be
reproduced in any manner or from without permission.